HOW TO BUILD & POWER TUNE

WEBER &
DELLORTO

DCOE, DCO/SP & DHLA CARBURETTORS

THIRD EDITION NOW IN COLOUR!

SpeedPro Series

4-Cylinder Engine - How to Blueprint & Build a Short Block for High Performance by Des Hammill
Alfa Romeo Twin Cam Engines - How to Power Tune by Jim Kartalamakis
BMC 998cc A-Series Engine - How to Power Tune by Des Hammill
BMC/Rover 1275cc A-Series Engines - How to Power Tune by Des Hammill
Camshafts - How to Choose & Time them for Maximum Power by Des Hammill
Cylinder Heads - How to Build, Modify & Power Tune Updated & Revised Edition by Peter Burgess
Distributor-type Ignition Systems - How to Build & Power Tune by Des Hammill
Fast Road Car - How to Plan and Build New Edition by Daniel Stapleton
Ford SOHC 'Pinto' & Sierra Cosworth DOHC Engines - How to Power Tune Updated & Enlarged Edition by Des Hammill
Ford V8 - How to Power Tune Small Block Engines by Des Hammill
Harley-Davidson Evolution Engines - How to Build & Power Tune by Des Hammill
Holley Carburetors - How to Build & Power Tune New Edition by Des Hammill
Jaguar XK Engines - How to Power Tune New Edition by Des Hammill
MG Midget & Austin-Healey Sprite - How to Power Tune Updated Edition by Daniel Stapleton
MGB 4-Cylinder Engine - How to Power Tune by Peter Burgess
MGB - How to Give your MGB V8 Power Updated & Revised Edition by Roger Williams
MGB, MGC & MGB V8 - How to Improve by Roger Williams
Mini Engines - How to Power Tune on a Small Budget 2nd Edition by Des Hammill
Motorsport - Getting Started in by SS Collins
Nitrous Oxide Systems - How to Build & Power Tune by Trevor Langfield
Rover V8 Engines - How to Power Tune by Des Hammill
Sportscar/Kitcar Suspension & Brakes - How to Build & Modify Enlarged & Updated 2nd Edition by Des Hammill
SU Carburettors - How to Build & Modify for High Performance by Des Hammill
Suzuki 4WD for Serious Offroad Action - Modifying by John Richardson
Tiger Avon Sportscar - How to Build Your Own Updated & Revised 2nd Edition by Jim Dudley
TR2, 3 & TR4 - How to Improve by Roger Williams
TR5, 250 & TR6 - How to Improve by Roger Williams
V8 Engine - How to Build a Short Block for High Performance by Des Hammill
Volkswagen Beetle Suspension, Brakes & Chassis - How to Modify for High Performance by James Hale
Volkswagen Bus Suspension, Brakes & Chassis - How to Modify for High Performance by James Hale
Weber DCOE, & Dellorto DHLA Carburetors - How to Build & Power Tune 3rd Edition by Des Hammill

Those were the days ... Series

Alpine Trials & Rallies 1910-1973 by Martin Pfunder
Austerity Motoring by Malcolm Bobbitt
Brighton National Speed Trials by Tony Gardiner
British Police Cars by Nick Walker
Crystal Palace by SS Collins
Dune Buggy Phenomenon by James Hale
More Dune Buggies by James Hale
Motor Racing at Brands Hatch in the Seventies by Chas Parker
Motor Racing at Goodwood in the Sixties by Tony Gardiner
Three Wheelers by Malcolm Bobbitt

Enthusiast's Restoration Manual Series

Citroen 2CV - How to Restore by Lindsay Porter
Classic Car Body Work - How to Restore by Martin Thaddeus
Classic Cars - How to Paint by Martin Thaddeus
Reliant Regal, How to Restore by Elvis Payne
Triumph TR2/3/3A - How to Restore by Roger Williams
Triumph TR4/4A - How to Restore by Roger Williams
Triumph TR5/250 & 6 - How to Restore by Roger Williams
Triumph TR7/8 - How to Restore by Roger Williams
Volkswagen Beetle - How to Restore by Jim Tyler

Essential Buyer's Guide Series

Alfa GT Buyer's Guide by Keith Booker
Alfa Romeo Giulia Spider Buyer's Guide by Keith Booker
Jaguar E-Type Buyer's Guide
Porsche 928 Buyer's Guide by David Hemmings
VW Beetle Buyer's Guide by Ken Cservenka & Richard Copping

Auto Graphics Series

Fiat & Abarth by Andrea & David Sparrow
Jaguar MkII by Andrea & David Sparrow
Lambretta LI by Andrea & David Sparrow

General

AC Two-litre Saloons & Buckland Sportscars by Leo Archibald
Alfa Romeo Berlinas (Saloons/Sedans) by John Tipler
Alfa Romeo Giulia Coupé GT & GTA by John Tipler
Alfa Tipo 33 Development, Racing & Chassis History by Ed McDonough
Anatomy of the Works Minis by Brian Moylan
Armstrong-Siddeley by Bill Smith
Autodrome by SS Collins & Gavin Ireland
Automotive A-Z, Lane's Dictionary of Automotive Terms by Keith Lane
Automotive Mascots by David Kay & Lynda Springate
Bentley Continental, Corniche and Azure by Martin Bennett
BMC's Competition Department Secrets by Stuart Turner, Phillip Young, Peter Browning, Marcus Chambers
BMW 5-Series by Marc Cranswick
BMW Z-Cars by James Taylor
British 250cc Racing Motorcycles by Chris Pereira

British Cars, The Complete Catalogue of, 1895-1975 by Culshaw & Horrobin
Bugatti Type 40 by Barrie Price
Bugatti 46/50 Updated Edition by Barrie Price
Bugatti 57 2nd Edition by Barrie Price
Caravans, The Illustrated History 1919-1959 by Andrew Jenkinson
Caravans, The Illustrated History from 1960 by Andrew Jenkinson
Chrysler 300 - America's Most Powerful Car 2nd Edition by Robert Ackerson
Citroën DS by Malcolm Bobbitt
Cobra - The Real Thing! by Trevor Legate
Cortina - Ford's Bestseller by Graham Robson
Coventry Climax Racing Engines by Des Hammill
Daimler SP250 'Dart' by Brian Long
Datsun 240, 260 & 280Z by Brian Long
Dune Buggy Files by James Hale
Dune Buggy Handbook by James Hale
Fiat & Abarth 124 Spider & Coupé by John Tipler
Fiat & Abarth 500 & 600 2nd edition by Malcolm Bobbitt
Ford F100/F150 Pick-up 1948-1996 by Robert Ackerson
Ford F150 1997-2005 by Robert Ackerson
Ford GT40 by Trevor Legate
Ford Model Y by Sam Roberts
Funky Mopeds by Richard Skelton
Honda NSX Supercar by Brian Long
Jaguar, The Rise of by Barrie Price
Jaguar XJ-S by Brian Long
Jeep CJ by Robert Ackerson
Jeep Wrangler by Robert Ackerson
Karmann-Ghia Coupé & Convertible by Malcolm Bobbitt
Land Rover, The Half-Ton Military by Mark Cook
Lea-Francis Story, The by Barrie Price
Lexus Story, The by Brian Long
Lola - The Illustrated History (1957-1977) by John Starkey
Lola - All The Sports Racing & Single-Seater Racing Cars 1978-1997 by John Starkey
Lola T70 - The Racing History & Individual Chassis Record 3rd Edition by John Starkey
Lotus 49 by Michael Oliver
Marketingmobiles, The Wonderful Wacky World of, by James Hale
Mazda MX-5/Miata 1.6 Enthusiast's Workshop Manual by Rod Grainger & Pete Shoemark
Mazda MX-5/Miata 1.8 Enthusiast's Workshop Manual by Rod Grainger & Pete Shoemark
Mazda MX-5 (& Eunos Roadster) - The World's Favourite Sportscar by Brian Long
Mazda MX-5 Miata Roadster by Brian Long
MGA by John Price Williams
MGB & MGB GT - Expert Guide (Auto-Doc Series) by Roger Williams
Micro Caravans by Andrew Jenkinson
Mini Cooper - The Real Thing! by John Tipler
Mitsubishi Lancer Evo by Brian Long
Motor Racing Reflections by Anthony Carter
Motorhomes, The Illustrated History by Andrew Jenkinson
Motorsport in colour, 1950s by Martyn Wainwright
MR2 - Toyota's mid-engined Sports Car by Brian Long
Nissan 300ZX & 350Z - The Z-Car Story by Brian Long
Pass Your Theory & Practical Driving Tests by Clive Gibson & Gavin Hoole
Pontiac Firebird by Marc Cranswick
Porsche Boxster by Brian Long
Porsche 356 by Brian Long
Porsche 911 Carrera by Tony Corlett
Porsche 911R, RS & RSR, 4th Edition by John Starkey
Porsche 911 - The Definitive History 1963-1971 by Brian Long
Porsche 911 - The Definitive History 1971-1977 by Brian Long
Porsche 911 - The Definitive History 1977-1987 by Brian Long
Porsche 911 - The Definitive History 1987-1997 by Brian Long
Porsche 911 - The Definitive History 1997-2004 by Brian Long
Porsche 911SC 'Super Carrera' by Adrian Streather
Porsche 914 & 914-6 by Brian Long
Porsche 924 by Brian Long
Porsche 933 'King of Porsche' by Adrian Streather
Porsche 944 by Brian Long
RAC Rally Action! by Tony Gardiner
Rolls-Royce Silver Shadow/Bentley T Series Corniche & Camargue Revised & Enlarged Edition by Malcolm Bobbitt
Rolls-Royce Silver Spirit, Silver Spur & Bentley Mulsanne 2nd Edition by Malcolm Bobbitt
Rolls-Royce Silver Wraith, Dawn & Cloud/Bentley MkVI, R & S Series by Martyn Nutland
RX-7 - Mazda's Rotary Engine Sportscar (Updated & Revised New Edition) by Brian Long
Singer Story: Cars, Commercial Vehicles, Bicycles & Motorcycles by Kevin Atkinson
Subaru Impreza by Brian Long
Taxi! The Story of the 'London' Taxicab by Malcolm Bobbitt
Triumph Motorcycles & the Meriden Factory by Hughie Hancox
Triumph Speed Twin & Thunderbird Bible by Harry Woolridge
Triumph Tiger Cub Bible by Mike Estall
Triumph Trophy Bible by Harry Woolridge
Triumph TR6 by William Kimberley
Turner's Triumphs, Edward Turner & his Triumph Motorcycles by Jeff Clew
Velocette Motorcycles - MSS to Thruxton Updated & Revised Edition by Rod Burris
Volkswagen Bus or Van to Camper, How to Convert by Lindsay Porter
Volkswagens of the World by Simon Glen
VW Beetle Cabriolet by Malcolm Bobbitt
VW Beetle - The Car of the 20th Century by Richard Copping
VW Bus, Camper, Van, Pickup by Malcolm Bobbitt
VW - The air-cooled era by Richard Copping
Works Rally Mechanic by Brian Moylan

First published in 1996 by Veloce Publishing, 33 Trinity Street, Dorchester DT1 1TT, England. Reprinted 1998. Second edition printed 1999 and reprinted 2000, 2001. Third edition first published 2002, reprinted 2003. This colour edition published 2005.
Fax 01305 268864/e-mail info@veloce.co.uk/web www.veloce.co.uk or www.velocebooks.com
ISBN 1-903706-75-0/UPC 6-36847-00275-6
Readers with ideas for automotive books, or books on other transport or related hobby subjects, are invited to write to the editorial director of Veloce Publishing at the above address.
British Library Cataloguing in Publication Data -
A catalogue record for this book is available from the British Library.
Typesetting (Soutane), design and page make-up all by Veloce on Apple Mac.
Printed in India.

HOW TO BUILD & POWER TUNE
WEBER & DELLORTO
DCOE, DCO/SP & DHLA CARBURETTORS

Des Hammill

THIRD EDITION NOW IN COLOUR!

VELOCE PUBLISHING
THE PUBLISHER OF FINE AUTOMOTIVE BOOKS

Contents

Introduction & Acknowledgements 7
Weber DCOE versus Dellorto DHLA 8
Acknowledgements..........................13

Essential information & Using this book 14
Essential information.......................14
Using this book...............................14

Chapter 1. Know your carburettor: stripdown & inspection 15
Know your carburettor - major components.....................................15
Chokes and auxiliary venturis 15
Idle mixture adjusting screws...........16
Main jet, emulsion tube & air corrector.......................................17
Progression hole inspection covers..18
Idle jet 19
Accelerator pump jets 19
Accelerator pump 20
Floats..21
Needle & seat..................................21
Stripdown - general advice..............22
Cleaning components 22
Fuel enrichment devices - special note...23

Stripdown procedure 23
Fuel enrichment device - blocking off discharge holes (Weber only) 28
Inspecting components for wear and damage...................................28
Weber & Dellorto 28
Weber only....................................29
Dellorto only.................................29
Difficult procedures 29
Clearing passageways 29
Throttle butterfly, spindle & bearings - maintenance, removal & refitting 29
Removing damaged threaded components.................................36
Removing jammed chokes and auxiliary venturis.........................37
Recognising 40mm emission controlled sidedraught Weber and Dellorto carburettors 38

Chapter 2. Rebuilding...................41
Needle valve & seat (Dellorto).........41
Floats & fulcrum pin (Dellorto)........41
Fulcrum pin - checking....................41
Floats - checking 41
Floats & pin - fitting 42

Float level - setting (Dellorto)..........42
Top cover (Dellorto) 43
Body components (Dellorto).............44
Needle valve & seat (Weber) 48
Floats & fulcrum pin (Weber) 48
Fulcrum pin - checking....................48
Floats - checking 48
Floats and fulcrum pin - fitting........48
Float level - setting (Weber)............49
Late model (mid 1980s-on) Spanish built Webers with plastic floats......49
Top cover (Weber) 51
Body components (Weber).............53

Chapter 3. Fuel management, air filters & ram tubes 59
Fuel filters 59
Fuel lines (pipes) & fittings 59
Air filters 60
Ram tubes (stacks)..........................61
Fuel pressure 62

Chapter 4. Choosing the components for your carburettor/s...............................64
Components - initial selection 64
Choke size versus carburettor size ...64

Choke size - selecting 65
Idle jet - selecting 67
Idle jet codes (Weber) 68
Emission controlled 40mm
   sidedraught Weber carburettors ..68
Idle jet codes (Dellorto) 68
Idle jet and air bleed component -
   selection 70
Idle mixture and progression holes ..70
Idle screw adjustment procedure 71
Main jet - selection 71
Emulsion tube - selection 71
Air corrector - selection 74
Auxiliary venturi - selection 75
Accelerator pump jet - selection 77
Accelerator pump intake/discharge
   valve 78
Needle valve - selection 78

Chapter 5. Manifold preparation &
   carburettor fitting **79**
Intake manifold - checking &
   preparation 79
The importance of stud alignment ...79
Carburettor - checking fit 80
Anti-vibration mountings 81
Carburettor/s - fitting to manifold 82

Chapter 6. Testing & set-up **85**
Idle speed 85
Fuel level & needle valve operation
   - checking 85
Throttle - initial adjustment and
   synchronization 86
Throttle arm fit...............................86
Throttle - initial setting (single
   carburettor) 86
Throttle - initial set-up &
   synchronization (multiple
   carburettors) 86

Idle mixture - initial adjustment 86
Throttle (linked throttle arm type)
   - final synchronization 89
Throttle (bar & pushrod type) - final
   synchronization............................91
Full throttle check 93
Idle jet alteration (fuel component) ..93
Ignition timing - general..................93
Idle jets/air bleeds - final selection....94
   Weber 96
   Dellorto 96
Idle mixture screws - final setting 98
'Idle by-pass' circuitry carburettors ..99
Adjusting idle by-pass systems.......100
CO readings and air'fuel ratio
   readings....................................101
Accelerator pump jets - final
   selection 101

Chapter 7. Rolling road tuning &
   problem solving **103**
Rolling road (dyno) testing
   procedure 103
Air/fuel mixture ratios....................106
Track testing procedure.................108
Solving problems - low to mid-range
   rpm 109
Solving problems - high rpm 109
Weber - fuel leakage from fuel
enrichment device..........................110
Weber - adapting for off-road
   applications 110
Return springs...............................111
Maintenance.................................111

Chapter 8. Fuel & octane
   ratings **112**
Heads without hardened exhaust
   valve seats 115
Fitting hardened valve seats 116

Cylinder heads with hardened
   exhaust valve seats 117
Sticking valves 117

Chapter 9. Jetting/setting
   examples **118**
BMC/Rover A-Series 1275cc
   engine (Weber).........................118
Ford 'Pinto' 2000cc SOHC
   standard engine (Weber)...........119
Ford 'Pinto' 2000cc SOHC
   standard engine (Dellorto).........119
Vauxhall 2000cc 16v engine
   (Dellorto) 119
Ford RS 2000cc SOHC Escort
   (Weber) 120
Ford Sierra Cosworth 2000cc
   (naturally aspirated)(Weber).......120
BMC B-Series 1900cc MGB
   engine (Weber).........................120
Toyota 4A-GE 1600cc 16-valve
   MR2 engine (Weber) 120
MG 1940cc alloy 8 port Magnette
   engine (Weber).........................120
Ford 1600cc Crossflow engine
   (Weber) 121
Jaguar XK 3.8 litre engine
   (Dellorto) 121
Ford 'Pinto' 2100cc engine fitted
   to a Formula 27 sports car
   (Dellorto) 121
Ford Sierra Cosworth 2000cc
   racing engine (Dellorto).............121
Ford 1760cc Crossflow engine
   (Dellorto) 121
BMC/Rover 1275cc A-series
   engine (Dellorto)......................122

Index..**127**

Veloce SpeedPro books -

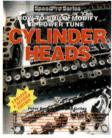
ISBN 1 903706 76 9

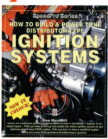
ISBN 1 903706 91 2

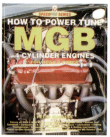
ISBN 1 903706 77 7

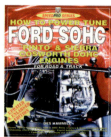
ISBN 1 903706 78 5

ISBN 1 901295 73 7

ISBN 1 903706 75 0

ISBN 1 901295 62 1

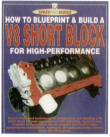
ISBN 1 874105 70 7

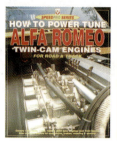
ISBN 1 903706 60 2

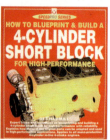
ISBN 1 903706 92 0

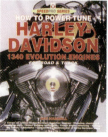
ISBN 1 903706 94 7

ISBN 1 901295 26 5

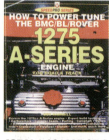
ISBN 1 901295 07 9

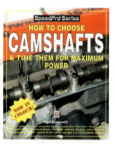
ISBN 1 903706 59 9

ISBN 1 903706 73 4

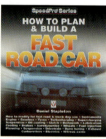
ISBN 1 904788 78-5

ISBN 1 901295 76 1

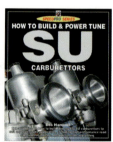
ISBN 1 903706 98 X

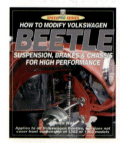
ISBN 1 903706 99 8

ISBN 1 84584 005 4

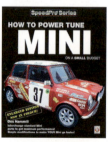
ISBN 1-904788-84-X

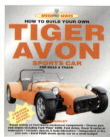
ISBN 1-904788-22-X

ISBN 1 903706 17 3

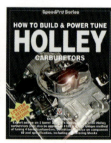
ISBN 1 84584 006 2

ISBN 1 903706 80 7

ISBN 1 903706 68 8

ISBN 1 903706 14 9

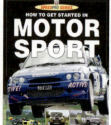
ISBN 1 903706 70 X

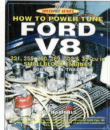
ISBN 1 903706 72 6

- more
on the
way!

Introduction & acknowledgements

The objective of this new book is to provide understandable information that will allow anyone with reasonable mechanical knowledge and aptitude to successfully strip, rebuild and tune Weber and Dellorto sidedraught carburettors for optimum performance.

There is no need to put up with a modified engine fitted with these fantastic carburettors that coughs and splutters or consumes huge amounts of fuel. The information in this book will help you to quickly isolate particular problems and alter the carburettor/s to suit your engine's actual requirements.

This completely revised book contains no details of choke (as in cold start device) settings because in most cases choke is never used with these carburettors. Cold starting is usually preceded by a partial pump of the throttle pedal to activate the accelerator pumps and, once the engine has actually started, working the accelerator pedal to keep the engine going until it is able to idle. Chokes just don't seem to get used because it's so easy to use the excellent accelerator pump system fitted to these carburettors instead.

Webers and Dellortos are truly excellent carburettors to work with.

A pair of Dellortos for an in-line four-cylinder engine.

Even in this day and age of high technology there still is a place for 'simple' carburettors that give high performance while basically being bolt-on items. Webers and Dellortos meet these requirements and their future is assured. One thing is for sure, on a cost-for-cost basis, Webers and Dellortos (especially if good second-hand carburettors are purchased) can give unrivalled value for money.

There is an old adage "what looks right is right" and this certainly applies to Webers and Dellortos. They always "look the part" on an engine because they are the part and, no matter what the engine size or type, they can be made to go as well as they look.

Although it may appear that each and every type and model of engine, and degree of modification, will require unique carburettor settings, this is not the case. It's very often the case that similar engines require quite similar jetting, and very good approximations can be made by experienced mechanics without even seeing the engine. It is a fact that engines can be categorized to quite a large degree, which is why it's been possible to narrow down the Weber and Dellorto components listed in this book to those you are likely to need and use. This will save you time and money.

I hope that you find this book informative and a practical help in the quest to tune these carburettors to get the best possible performance from your car, with reasonable economy.

Weber DCOE versus Dellorto DHLA

Argument has raged for years about whether Weber or Dellorto carbs are better. No realistic comparison is possible, however, unless all of the available adjustments have really been optimized on each carburettor type and on the same engine.

40 DHLA Dellorto.

45 DCOE Weber.

Fortunately, both carburettors are so good it doesn't really matter which manufacturer you choose. The main thing everyone needs to know is that you can buy either with absolute confidence in their performance, and that both can be tuned equally well. There are no really bad sidedraught Webers or Dellortos, but some are better than others. Both manufacturers have made emission versions of their respective carburettors, but these are not as good as the 'universal performance' versions, in terms of best possible all round accelerative and top end performance. When using secondhand Weber or Dellorto sidedraught carburettors you do need to know what you're buying by being able to identify exactly what is on offer. No competition engine should ever be equipped with emission type Dellorto or Weber 40mm carburettors.

Note that one of the quickest ways to establish at a glance whether a sidedraught Weber or Dellorto carburettor is suitable for high-performance use is to count the number of progression holes. Any carburettor which has two or three very small diameter (1mm/0.040inch) progression holes will have a rich progression phase, all other factors being equal. Carburettors with five quite large diameter (2mm/0.080inch) progression holes will have a weaker progression phase, all other factors being equal.

Nothing of any consequence interchanges between the two makes of carburettor. The only thing common to both is that they will bolt on to the same intake manifold.

The design differences between the two carburettor makes show that their respective manufacturers have achieved the same objectives, but by different means. An example of this is the accelerator pump action. In both

40 DHLA Dellorto.

45 DCOE Weber.

Contents of a non-genuine aftermarket Dellorto repair kit for two carburettors.

The genuine Dellorto repair kit for two carburettors is more comprehensive than aftermarket ones.

and built in the days prior to reliable fuel resistant plastics and originally had brass floats, for instance.

The sidedraught Weber was undoubtedly the original in its DCO form of the early 1950s, and the overall concept and design the work of a genius (Edoardo Weber). The DCO series of sidedraught carburettors was superseded in the 1960s by the more affordable die cast sidedraught DCOE.

Norman Seeney Ltd. (Tel: +44 (0)1527 892650, Fax: +44 (0)1527 893017) repairs and maintains the early sand cast DCO3 and DCOA3 series carburettors, as fitted to Aston Martin, Ferrari, Maserati, Jaguar and Coventry Climax engines, for example. In fact, the company offers a worldwide repair and maintenance service for all early Weber racing carburettors. The most regular types of carburettor Norman Seeney services and repairs are: 38mm-40mm DCO3, 42mm, 45mm, 48mm, 50mm and 58mm DCO3, 58mm DCOA3 (six bolt flange), 40mm, 42mm, 45mm DCOA3 (four bolt flange). Others include the 48mm DOM, the 52mm DCO and the 36mm DOE.

There is no doubt that the racing success rate of cars equipped with these original sidedraught carburettors, and the fact that Dellorto didn't start making sidedraught carburettors until the late 1960s, contributed to Weber having the best known 'name'. Second-hand prices are generally higher for Webers even though the new price of each was similar. There are literally thousands of both types of sidedraught carburettor scattered around the world and millions of tuning parts sitting in tool boxes and garage cupboards. These carburettors are going to be around for a very long time and continue to be fitted to a wide variety of engines.

The Dellorto company designed

cases the engine receives an identical amount of fuel but one carburettor has a diaphragm and the other a plunger. The Dellorto exhibits what can be termed modern manufacturing techniques which modernize the original Weber carburettor concept to a small degree. The Weber was designed

Triple DCO3s on this D-Type Jaguar engine.

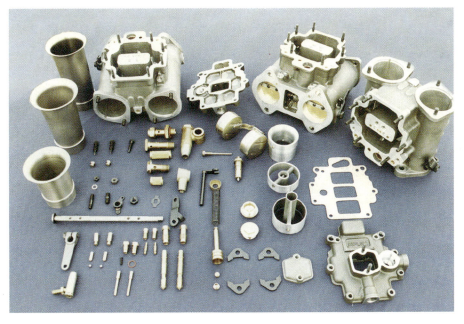

Just some of the components available from Norman Seeney Ltd.

mixture leakage problems).

The Weber choke/fuel enrichment device can be troublesome when it gets a bit worn and can pass a huge amount of extra air/fuel mixture into the engine; in fact to such an extent that the engine may not even run. The problem being that, although the mechanism is in the off position, it only takes a bit of lever wear and a jammed starter valve (in the up position) and you have a very rich mixture. However, if the choke has never been used (most have not) this situation will not arise. The choke outlets can be blocked off permanently to preclude this happening. Note that if either of the starter valves found in a carburettor do not seal off they will allow fuel mixture to pass by.

The venturi choke and auxiliary venturi are held in position with a single taper point screw and lock nut on all Dellorto carburettors and this is an excellent retention method. The Weber uses a single taper point screw for each on 45s and 48s with a securing plate linking the two screws and bent tabs for a locking arrangement. The locking tabs must always be fitted to prevent the screws from winding out (they can be lockwired).

The DCOE 40s on the other hand have blade spring location on both the auxiliary venturi and the choke. Both are held in position by the trumpet and its two retaining clamps. On well used carburettors the trumpets can often rotate, even when the trumpet clamps are fully tightened. There are several reasons for this (none of them good) but the end result is that the carburettor body gets worn and even a new trumpet, choke and auxiliary venturi will not restore the situation although specially made trumpet clamps (stepped) will cure the problem.

The accelerator pump lever

their carburettor with the view to improving the Weber formula and they have achieved this to a certain extent. The integral, plastic-caged fuel filter is well done and better than the soldered brass mesh tube of the original Weber or the plastic-caged filter alternative available these days from Weber

(which often crushes the first time it is installed). The choke operation is certainly better on the Dellorto (same overall principle) from the standpoint of having a good shut off by way of the neoprene washer and the piston action that uncovers the holes for the fuel /air mixture to pass through (no fuel

Unusual twin sidedraft application. Two Weber carburettors, but only one choke from each is used. Note how areas of unused chokes have been cutaway to provide clearance.

is not as compact on the Dellorto as the Weber's (which it is totally enclosed within the carburettor body). The Dellorto components could be damaged through lack of care but it is not usual for this to happen: in fact, both carburettors are pretty hardy units. The more modern style of the Dellorto's construction and methods it uses to duplicate the Weber principal of operation have been very successful overall. The differences between the carburettors mentioned here are all pretty minor in the overall scheme of

things but worth noting to illustrate that both carburettors have strengths and weaknesses when compared to each other. Dellortos are a little bit easier to service. The Weber is more compact height-wise.

The method of tuning either carburettor is essentially the same. Unfortunately, in both cases, the axiom "bigger is better" seems to prevail but, in most instances, this is wrong and most engines that do not run well and prove to have carburettor problems have chokes that are too large, main jets that are too large, accelerator pump discharge jets that are too large or too much fuel pressure. It is difficult to understand why anybody would want to go to the time and trouble of fitting Webers or Dellortos and for the sake of a comparatively small amount of money put up with an uneconomical and poorly performing engine. This happens all the time and the carburettors are often blamed for it when, in fact, this is not possible as both carburettors are infinitely tunable to achieve perfection in all instances. It is the tuning of the carburettors that is the real problem.

The competition engine built

professionally and on a big budget is run, tuned and tested by the particular concern doing the job. However, the majority of Weber and Dellorto users do not have the sophisticated equipment that a major tuning company will have, yet their car's engine still needs to be tuned correctly. In fact this is not such a big problem because the principles of Weber and Dellorto carburettors are logical and understandable so that, when sound tuning techniques are employed, both carbs can be tuned correctly with a minimum of equipment.

The average enthusiast will never have the same resources at their disposal as the professional but, with care and attention to detail, can get an engine tuned equally well. Of course, money can be saved if the right choice of components is made first time! All of the jet sizes given in this book are approximations because individual engines vary so much. If your engine is being rebuilt and you intend to use high revs, make sure it is built with the biggest permissible tolerance sizes on the piston to bore clearance and the main and big end bearing clearances (factory specifications, but largest sizes permissible). It is not possible to tune an engine that has some mechanical problem. Well-built engines respond perfectly to correctly tuned sidedraughts.

Note that Dellorto and Weber have both supplied their carburettors to car manufacturers, such as Alfa Romeo, as original equipment. These application specific 'emission' type sidedraught carburettors tend to cost less on the secondhand market than 'universal performance' versions because of what can be their limited 'tune-ability' when fitted on to other engines. The more desirable 'universal performance' 40mm DHLA40 carburettor (no suffix letter), comes

A pair of Dellortos on the Vauxhall engine of a kit car.

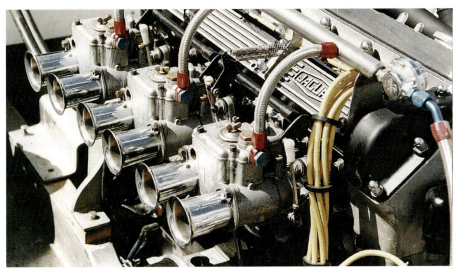

A trio of Webers on a racing Jaguar engine.

in two models, one with three small diameter progression holes (1.0mm), the other having four. A visual check is the only way to tell them apart. For racing purposes, the three progression hole carburettor will usually give the best performance. The DHLA40E has four slightly larger diameter progression holes, and the DHLA40C has six small diameter progression holes. What makes these four Dellorto carburettors different from each other on the basis of performance characteristics is the number and size of the progression holes. With 40mm Weber sidedraught carburettors the DCOE 11 and the DCOE 2 models are the ones to have.

The Weber and Dellorto emission type carburettors feature idle mixture adjustment screws in towers, have a vacuum take off on one carburettor, very small or no float chamber vent holes in the carburettor body on some Webers, and vented jet inspection covers. None of these features actually affect the tune ability of the carburettors, but the design criteria and differences inside them does. The emission 40mm Dellortos, for example, have factory drilled holes in the carburettor body that control the air bleed for the idle mixture and the progression phase mixture strength. While this is excellent for low rpm use emission control, it's not ideal for when all out maximum accelerative engine performance is required.

This book differentiates between 40mm 'emission' Weber and Dellorto carburettors and 'universal performance' 40mm Weber and 40mm Dellorto carburettors on the basis of the available engine performance with each type of carburettor used. Get and use the right model of carburettor for your application!

The advent of electronic fuel injection systems for production engines has not replaced the sidedraught Weber or Dellorto in the eyes of many, because these systems have brought in an element of complication and expense. The difference in overall performance between the fuel injection and well tuned sidedraughts can be very small, yet the cost between the sidedraughts and the up-rated fuel injection system can be quite large. In spite of modern fuel injection systems, Weber and Dellorto sidedraught carburettors are here to stay, and for a lot longer than you might imagine.

It used to be the case that new parts were more readily available for Webers than for Dellortos. This has changed now, and the master agents of both type of carburettor have access to all parts. The DCO/SP Weber carburettors are, as of 2005, not available new, although Webcon in the UK is now offering DCOW 45 and 47mm versions of the DCOE type carburettor. Dellorto hasn't made DHLA sidedraught carburettors for some years now but spare parts are readily available, if a bit expensive. Contact your country's main agent, who should be able to help you. Eurocarb in the UK, for example, will sell Dellorto and Weber parts to anyone anywhere in the world, and has all Dellorto DHLA parts in stock for immediate shipment. The Weber DCOE sidedraught stocks are also very comprehensive, and available for immediate shipment.

Acknowledgements

Thanks to Matthew Cooper of Eurocarb Ltd for his most valued assistance with spare parts for photographs and information. Thanks also to Steven Miles of The Tipton Garage, Tipton St. John, Nr. Sidmouth, Devon, England, Brian Wills of Kings Mews Racing, Newton Abbot, Devon, England, Paul Kynaston of Kynaston Auto Servies Ltd, Marsh Barton, Exeter, England, Andy Gray of Webcon and Norman Seeney.

With thanks, as always, for her support and assistance to my wife Alison.

Des Hammill

Essential information & using this book

ESSENTIAL INFORMATION

This book contains information on practical procedures; however, this information is intended only for those with the qualifications, experience, tools and facilities to carry out the work in safety and with appropriately high levels of skill. Although the words **Warning!** (personal danger) and **Caution!** (danger of mechanical damage) are used throughout this book, be aware that we cannot possibly foresee every possibility of danger in every circumstance. Whenever working on a car or component, remember that your personal safety must **ALWAYS** be your **FIRST** consideration. **The publisher, author, editors and retailer of this book cannot accept any responsibility for personal injury or mechanical damage which results from using this book, even if caused by errors or omissions in the information given. If this disclaimer is unacceptable to you, please return the pristine book to your retailer who will refund the purchase price.**

This book applies to all Weber DCOE series and Dellorto DHLA series sidedraft carburetors.

It is possible that changing carburetor specification will mean that your car no longer complies with exhaust emission control or other regulations in your state or country - check before you start work.

An increase in engine power and, therefore, performance, will mean that your car's braking and suspension systems will need to be kept in perfect condition and uprated as appropriate.

As these carburetors were built to metric measurements, these take priority in the text. It is essential that you work with metric wrenches, but we would also advise you to use metric measurements if you can.

The vital importance of cleaning a part before working on it cannot be overstressed. It is equally important that your tools and working area are completely clean.

Use good quality tools and make sure they are precisely the right size for every job.

USING THIS BOOK

You'll find it helpful to read the whole book before you start work or give instructions to your contractor. This is because a modification or change in specification in one area will cause the need for changes in other areas. Get the whole picture so that you can finalize specification and component requirements (as far as possible) before any work begins.

This book has been written in American English; those in any doubt over terminology will find a glossary of terms at the back of the book.

Chapter 1

Know your carburettor: stripdown & inspection

KNOW YOUR CARBURETTOR - MAJOR COMPONENTS

The overall layout is almost identical for Webers and Dellortos. The principles of operation are so close that carburettors of this type couldn't really be designed any differently.

Chokes and auxiliary venturis

On all carburettors the choke is fitted into the barrel bore first and then the auxiliary venturi.

Note: it is quite possible on the 45 and 48 DCOE to put the auxiliary venturi into the carburettor body the wrong way around. The engine will run quite satisfactorily at low rpm but will run lean as the revs rise. Always check to see that the auxiliary venturi is fitted the right way around.

View of a Weber showing the idle mixture adjustment screw at (A) the pump jet screw cover at (B) the throttle arm adjustment screw at (C) the progression hole screw cover at (D) the main jet and idle jet cover at (E).

View of a Dellorto showing the idle mixture adjusting screw at (A) the pump jet screw cover at (B) the throttle arm adjustment screw at (C) the progression hole screw cover at (D) the main jet and idle jet cover at (E).

Idle mixture adjusting screws

The idle mixture adjusting screws are in a similar position on both carburettors. There are two types of adjusting screws for each carburettor. One type, used by both Dellorto and Weber, has the spring and idle mixture adjusting screw totally in the open. This type of screw has a metric course thread.

The second type of mixture adjusting screw is enclosed in a tower which is part of the body casting. The Dellorto has a fine thread cut into the tower. These Dellorto idle adjustment screws are notorious for becoming jammed. The thread jams in the tower (due to the accumulation of dirt, dust and corrosion) and any turning pressure applied with a screwdriver ruins the slot gets ruined. The only course of action is to remove the carburettor, place it in a machine vice on a turret type milling machine

45 DCOE auxiliary venturis shown the right way round for installation into a carburetor - the discharge slot must face the butterfly. Note the difference between the two types of auxiliary venturi. The one on the right, which has a smaller venturi (7.5mm compared to 10mm), is better for high performance use.

(Bridgeport or similar), line the true centre of the adjustment screw up with the spindle of the machine and, using a milling cutter, mill the screwdriver slot/threaded portion of the adjustment screw away. Then, using a small pick, the thread that is left in the

Dellorto DHLA carburettors with the choke and auxiliary venturi in front. Note the shape of the auxiliary venturi, the 40's (top picture) is slightly different to the 45/48's (above).

Exposed type of idle mixture adjusting screw on a Weber. The Dellorto carburettors equipped with exposed idle mixture adjustment screws look very similar.

Weber 40 DCOE carburettors with the chokes and the auxiliary venturis shown in front. The shape of the 40's auxiliary venturi (top) is different to the 45/48's (above).

Idle mixture adjusting screw tower of a Weber carburettor. Dellorto carburettors equipped with towers look very similar and their idle screws have very fine threads.

carburettor body is 'pulled' out. This is work for a specialist. Weber threads (coarse metric) are cut into the body in the usual place, and an O-ring is positioned at the top of the screw to keep dirt out.

Main jet, emulsion tube & air corrector

These three components form a modular component which screws into the body of the carburettor as a single unit. The main jet, emulsion tube and air corrector can be removed from either carburettor through the top cover. The Weber carburettor has a wing nut with a round cover assembly which is easily removable and which gives access to the jets. The Dellorto has a plastic or aluminum cover held by two screws which, when removed, allows access to the jets.

The jets are a push fit into the emulsion tubes on both carburettors. The air corrector on the Weber is a

Dellorto very fine threaded idle mixture adjusting screw which fits in tower model carburettors on the left. Weber tower type idle mixture adjusting screw on the right.

Top of a Dellorto with the jet inspection cover removed and showing the emulsion tube holders which, in turn, hold the actual emulsion tubes, main jet and air corrector.

Below - Progression holes in the body of a Dellorto carburettor. These holes must be clear for good progression, though it's unusual for them to become blocked.

push fit into the emulsion tube and the emulsion tube holder is a push fit on to the emulsion tube which, in turn, covers the air corrector. This means that there are two parts involved, the actual air corrector jet and the emulsion tube holder.

On the Dellorto the air corrector is part of the emulsion tube holder and is changed as a single component. The emulsion tube holder/air corrector is a push fit on to the emulsion tube.

Progression hole inspection covers

This cover was put here in the first place to allow for the progression holes to be drilled in the carburettor body but, of course, it doubles as a removable plug for cleaning purposes or checking to see if the holes are clear. Note: some Weber carburettors have permanent plugs.

Weber main jet, emulsion tube, air corrector jet and an emulsion tube holder dismantled (right) and assembled (left).

Dellorto main jet, emulsion tube and an emulsion tube holder dismantled (right) and assembled (left). Note that the air correction hole is part of the holder.

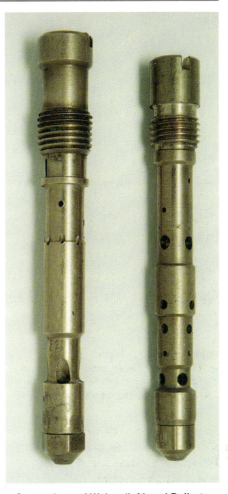

Comparison of Weber (left) and Dellorto (right) main jets, emulsion tubes and holders.

On well used Webers replace the return spring (fitted behind the inspection cover). The 'eyes' break off these springs after long service.

Idle jet
The idle jet holder and the idle jet are located under the same inspection cover that gives access to the main jet, emulsion tube and air corrector jet. The idle jets are in a similar position in both carburettors. The idle jet is a push fit into the idle jet holder which, in turn, screws into the carburettor body. The air bleed and fuel jet are in the same component on the Weber. On the Dellorto the fuel jet is a separate component to the air bleed which has one or more holes drilled into the actual idle jet holder.

Accelerator pump jets
These are found under the screw plug and are lifted out. On both carburettors they can sometimes be difficult to remove. **Caution!** Extreme care is often needed to remove them without damaging or even ruining them. On Dellortos the pump jet is actually connected to the screw plug using a

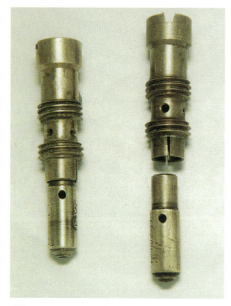

Weber idle jet and holder assembled (left) and separated (right).

Weber carburettor with cover removed showing the position of the idle jets.

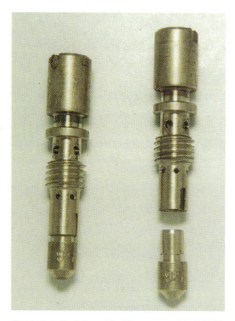

Dellorto idle jet and holder assembled (left) and separated (right).

socket and groove arrangement for ease of removal.

Accelerator pump

The Weber mechanism is totally enclosed in the carburettor top and

Dellorto carburettor with cover removed and showing the position of the idle jets.

body but there is also a cover at the back of the body where the throttle return spring and the accelerator pump control lever is situated.

The Dellorto has an external system with a diaphragm operated accelerator pump and lever system (all located underneath the carburettor).

Dellorto accelerator pump unit and operating mechanism.

Top cover turned upside-down to show the float arm in contact with the needle valve.

Needle and seat in assembled and component form.

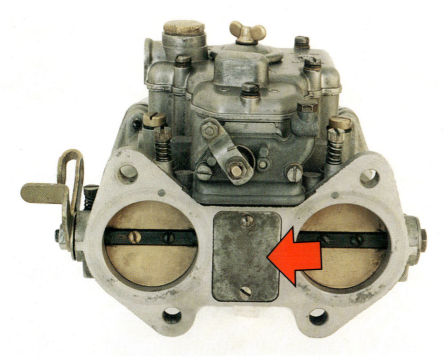

Weber carburettor from the back showing the inspection cover which serves to enclose the accelerator pump actuating mechanism.

carburettor the wing nut and round cover must be removed before the carburettor cover will come off. The floats are located on the underside of the carburettor cover and, as a consequence, the carburettor cover has to be removed to gain access to the floats for float height setting purposes. The Weber uses brass and, latterly plastic, floats and the Dellorto uses plastic floats only.

Needle & seat

These are situated in the top cover and work in conjunction with the floats. The float must be removed before the needle and seat can be removed.

Caution! Care must be exercised when removing the floats, and, more specifically, when removing the float fulcrum bars. It's possible to break the cover casting through rough handling and render the cover useless. It's also

Floats

The floats are located under the carburettor top cover on both carburettors. Note: on the Weber

This Dellorto carburettor top cover has had one of the legs broken off.

possible to damage or break off one of the split legs through careless removal. Always tap the bar through, using a pin punch, from the solid leg side and NOT the split leg side.

STRIPDOWN - GENERAL ADVICE

Both Weber and Dellorto carburettors can only perform to design specification if all component parts are in good condition. These carburettors are nearly always able to be fully rebuilt. Only severe damage will cause a carburettor body to be rejected while the majority of the other parts could still be used for spares. Carburettor rebuilding applies to used carburettors but new carburettors may have to have their settings checked if they malfunction (this is rare).

Both Weber and Dellorto carburettors are, within reason, infinitely re-buildable. Carburettors covered with road grime are not necessarily in poor condition internally. Rough handling and general abuse

are what usually cause functionality problems with these carburettors. Poor examples usually feature burred or broken screwdriver slots, burred hexagon fittings and generally look knocked around. All of the parts are replaceable and the only thing that precludes a carburettor from a rebuild is a seriously damaged body. It is possible to weld certain parts on a body but this is not usually cost-effective so damaged bodies generally have to be written off.

When rebuilding either of these carburettors it is useful to refer to an exploded diagram of the carburettor. The exploded view diagrams included in this book have been kindly supplied by Dellorto and Weber main agents. In addition, another way of ensuring that you have a reference from which to check exactly where, or which way around, a part or parts go, is to dismantle only one carburettor at a time. (If you only have one carburettor, place the parts in line in their order of removal). The exploded

diagrams included in this book are very good and all parts are shown in order of fitting. Some of the parts are not shown directly in line with their actual fitted position in the carburettor body, but their positions are correct in the overall scheme of installation.

Cleaning components
The following will be required for cleaning the carburettors; two shallow pans, a 1 inch or 25mm paint brush, half a gallon or 2 liters of gasoline (petrol) and pint or 750ml of paint thinner and, perhaps, some specific carburettor cleaner as found and supplied in spray cans (300mls). Warning! These cleaning agents are highly inflammable and may be harmful to skin, or to the lungs if inhaled - always take appropriate precautions to ensure your complete safety.

In the first instance wash and clean the outside of the carburettor thoroughly with a degreasing agent such as petrol so that all loose dirt and grime is removed. It is not desirable to immerse the carburettor completely during the initial washing as dirt can be washed into the carburettor. Once the majority of dirt has been removed the outside surfaces can be cleaned with paint thinners or a proprietary carburettor degreasing agent (usually available in a spray can) which will clean the outside surfaces as well as is possible without full immersion. The outside surfaces of the carburettors do get quite badly stained with use and there is usually slight surface corrosion which can't really be removed without using a proprietary aluminum cleaning agent. **Warning!** If you do use such an agent make sure it is appropriate for this task and follow the maker's instructions fully.

One tray can be used for holding the petrol and washing down the carburettor and parts and the other for keeping the parts together once removed from the carburettor and cleaned . When the majority of the dirt and grime has been removed from the carburettor and its components, change the fluid or use another tray to give all parts a final wash so that they are completely clean.

Caution! Never clean jets or passages with wire: use compressed air, nylon bristles or a piece of appropriately thick fishing line for this purpose.

Fuel enrichment devices - special note

Because the fuel enrichment devices are never used, no details of their stripping down or assembly are given. Leave all of the fuel enrichment components installed in the carburettor. The Weber and Dellorto fuel enrichment devices work perfectly, of course, but never seem to get used. Most engines equipped with these carburettors that are well tuned need one pump (at the most) for starting and some throttle 'feathering' to keep them running for the first 10 to 20 seconds after cold start-up and then the engine will usually idle on its own account. This idle may not necessarily

be up to the set idle speed while the engine is cold but will improve as the engine gets warm.

On the Weber carburettor the fuel enrichment device frequently causes problems (excessive fuel delivery), so, rather than repair the mechanism, simply block off the outlet holes in the throttle bores. This is done by tapping the holes and screwing in a grub screw (use a locking agent and peen over the hole to prevent the grub screw unwinding and going into the engine), so that fuel cannot enter the engine via the enrichment holes. The Dellorto will only allow an over-enriched fuel supply to enter the engine if the fuel enrichment device washer fails, which it seldom does.

STRIPDOWN PROCEDURE

The carburettor can now be stripped. Remove the fuel pipe union using a six-point box end wrench (ring spanner). There is not all that much hexagon to grip onto (depth-wise) on the union bolt and it is fairly usual for the hexagon to be burred. (If the carburettors are on a manifold it is usual to loosen off the union bolt and the filter inspection bolt [on Webers] before removing a carburettor as they can often be very tight - this way there

is more to hold firmly while the bolt is actually undone).

The fuel filter can now be removed and, in the case of the Weber, expect to have to replace it with a new item. The Dellorto filter can be cleaned and inspected for any damage to the screen and, if clear, re-used. Note: with Weber carburettors, the internal fuel filter can be left out provided there is an inline fuel filter in the system between the fuel pump and the carburettors. The Weber filter is not as good as the Dellorto and, even if the standard filters are left in place, an additional filter should be used as a precaution against dirt getting into the carburettor. Expect to fit new fibre washers to the union bolt and union, plus one for the filter inspection plug on Webers.

Remove the jet inspection covers (wing nut on Webers and two screws on Dellortos) and, using a correctly-sized screwdriver, remove the main jets and idle jets. Pull the components apart for thorough cleaning. Check each part for corrosion and use 320 wet and dry paper to clean off any stain. Damaged screwdriver slots can be cleaned up using a rectangular section needle file while holding the emulsion tube holder or idle jet holder firmly in a vice between pieces of wood (to protect the holder). Ideally, use compressed air to clean each component after washing it and check visually that each jetting hole is clear and looks clean.

Remove the accelerator pump screw plugs and lift out the pump jets. The Dellorto pump jet is designed to come out with the screw as the screw has an internal groove in it and the jet itself has a radius-edged lip at the top which fits into the screw plug. Normally this works very well, but if the pump jet has not been removed for a long time, it may be jammed in the hole in which

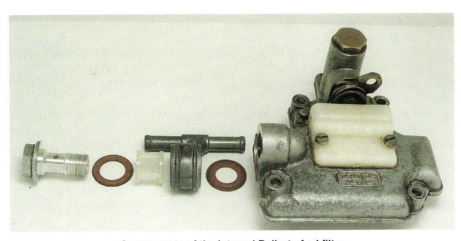

Components of the integral Dellorto fuel filter.

case the screw plug parts company with the jet. The Weber screw plug simply lifts off and the pump jet can usually be removed by using pointed tweezers.

On both carburettors, if the pump jet is well and truly jammed in the hole a pair of long-nosed pliers can be modified (by grinding) so that the pump jet can be gripped and pulled out. The Weber pump jet has a small groove around the top edge for this very thing and the Dellorto has a radius-edged lip which can be used for the jet's extraction.

Remove the progression hole inspection screw plugs and check to see that there is no obvious dirt covering the holes. Use the correct sized screwdriver for the slot size.

Remove the idle mixture adjusting screws and check the tapered ends for any sign of damage: they should all look the same. It is unusual to find a carburettor or a pair of carburettors with all screw ends damaged so compare all screws to each other.

Some Dellortos have a tower and the idle mixture adjusting screw is located within this tower. A fine thread is cut into the tower and, because of this design, the threads are affected by dirt collecting in the recess of the tower. The idle mixture adjusting screws can be quite tight to move, initially at least. Usually, once the carburettor body has been thoroughly washed and cleaned with compressed air, the idle mixture adjusting screws can be refitted without any problem. These threads can, unfortunately, become seriously damaged and be very difficult to repair. **Note:** there is something that can be done to prevent any damage occurring to the threads in the future. The idle mixture adjusting screws are not used often so the solution to this potential problem is to plug the holes after the engine

has been tuned. The simplest way is to use silicone sealer. The plug of sealer can easily be removed when an adjustment needs to be made and new sealer applied afterwards. The other way of doing this is to fit small plastic plugs that can be pushed in and then removed for idle mixture adjustment when necessary. Caution! If the engine is always getting dirty because of how the car is used, it is essential to seal the towers off.

If you cannot move the idle mixture screw, the screw's head has to be ground away. This is done with a high speed grinder fitted with a small ball-nosed cutter. Care must be taken so that the internal thread of the tower is not damaged any further. No attempt should be made to turn the screw out. Grind it all away and remove all traces of the screw material from the threads. Sometimes it is possible to do this and simply fit a new idle mixture adjusting screw, especially if a damaged screwdriver slot is the only reason for not being able to undo the screw. With the head removed the remaining portion of the screw is loose. Remove all metal particles.

Undo the screws from the top cover and, very carefully, remove it without any forcing of the floats to avoid damaging them. The Dellorto has a particularly close fit around the choke tower which necessitates that it be lifted vertically for a few millimeters. Turn the top over and, using a small pin punch, very carefully remove the fulcrum pin that holds the floats in position (**Caution!** - it's easy to break the split posts). Push the fulcrum pin out from the solid leg side, not the split leg side. Refit the fulcrum pin from the solid leg side too. There is a gasket between the carburettor body and the top (on both carburettors) which can only be removed and replaced while the floats are not fitted. It is always a

WEBER DCOE - EXPLODED VIEW
1 Filter inspection cover, 2 Fiber washer for inspection cover, 3 Fuel filter, 4 Main jet and idle jet inspection cover, 5 Carburettor cover screw, 6 Inspection cover gasket, 7 Gasket for carburettor cover, 8 Needle and seat, 9 Plastic floats, 10 Emulsion tube holder, 11 Air corrector, 12 Idle jet holder, 13 Emulsion tube, 14 Stud, 15 Main jet, 16 Idle jet, 17 Auxiliary venturi, 18 Trumpet or ram tube, 19 Nut, 20 Spring washer, 21 Retaining plate, 22 Choke, 23 Stud, 24 Retaining screw locking tab, 25 Cap, 26 Adjusting screw, 27 Locking nut, 28 Auxiliary venturi retaining screw, 31 Locking washer, 32 Spindle retaining nut, 33 Butterfly, 35 Butterfly retaining screw, 36 Bottom well gasket, 39 Inspection cover retaining screw, 40 Choke mechanism retaining screw, 41 Inspection cover gasket, 42 Choke mechanism, 43 Accelerator pump intake and discharge valve, 44 O-ring, 45 Linked throttle arm (left-hand side), 46 Washer, 47 Idle mixture adjusting screw spring, 48 Progression holes inspection screw, 49 Accelerator pump aluminum washer, 50 Spindle return spring, 51 Spindle return spring anchoring plate, 52 Throttle arm adjusting screw, 53 Accelerator pump piston, 54 Starter valve, 55 Starter valve return spring, 56 Idle mixture adjusting screw, 57 Accelerator pump jet, 58 O-ring, 59 Pump jet screw plug, 60 Accelerator pump control rod spring, 61 Accelerator pump control rod, 62 Starter jet, 63 Check valve ball, 64 Stuffing ball, 65 Accelerator pump stuffing ball screw plug, 66 Aluminum spacer washer, 67 Small bore fibre washer, 68 Fuel union (banjo), 69 Large bore fibre washer, 70 Union bolt, 71 Progression holes cover, 72 Float fulcrum pin, 73 Choke cable retaining screw, 74 Starter jet holder, 75 Carburettor top cover.
(Courtesy Weber).

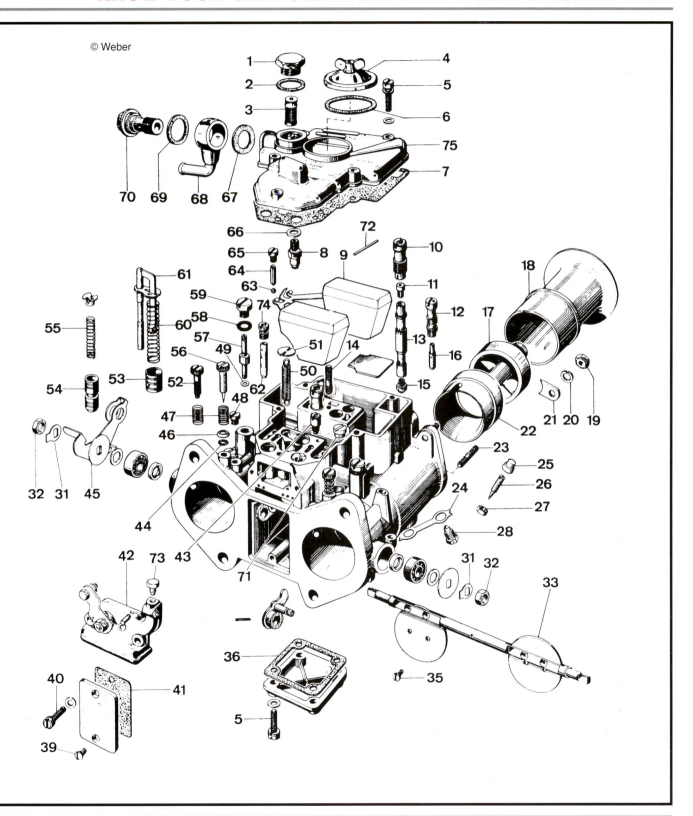

© Weber

good idea to replace this main gasket with a new one; essential if there is any sign of deterioration.

Shake the floats to make sure they contain no fuel and then check their integrity by immersing them completely in fuel and watching for air bubbles. Replace any faulty Dellorto float. Brass Weber floats can be repaired by soldering. Any soldered repair has to be reasonably small so that a minimum amount of weight is added to the float assembly. If the floats and frame look battered and well used, replace them with new components. All Dellorto floats are plastic and deteriorate over time where the aluminum of the frame has the plastic moulded around it. The floats' main function is to maintain a constant fuel level under all conditions by opening and closing the needle valve. Worn fulcrum pins and float hinges can cause fuel level to fluctuate.

The needle and seat assembly can be unscrewed and should be replaced with a new assembly unless you are working on a relatively new carburettor. The needles are prone to wear on the very tip of the needle where it contacts the seat. Worn needles have an annular groove or indentation present all the way around on the conical portion of the needle. The wear on the seat is less apparent as it is down the hole of the fitting. Replace the needle and seat as an assembly. Needles can be all metal or rubber tipped.

Remove the trumpets, auxiliary venturis and chokes. The Weber DCOE 40 has the choke and auxiliary venturi retained by the trumpet and clamps. Some models of the 40 DCOE have a different auxiliary venturi which allows an air box to be directly bolted on the carburettor without the usual clamps. The 45 and 48 DCOE Webers have a screw for each choke and auxiliary venturi and the Dellorto has

one nut and screw to remove on the auxiliary venturi before anything can be removed. All of these components usually come out easily but the chokes can be difficult to remove. If the chokes won't budge refer to the information on choke removal given in the "Difficult Procedures" section of this chapter.

Turn the carburettor over and remove the bottom well cover. The main jets take their fuel out of here and usually there is sediment and possibly some corrosion in the cover and the underneath cavity of the carburettor body. The sediment and corrosion must be washed and/or carefully scraped away.

On the Weber there are four screws holding the well cover on and there is a gasket between the cover and the body. Use a new gasket during the rebuild.

On the Dellorto there are two parts to the bottom cover arrangement. The accelerator pump housing is below the well cover and has to be undone first. The accelerator pump arm is part of the lower housing and, when the four screws are undone, the housing can be moved aside while the accelerator pump arm is all still attached. Do not move or undo the two nuts that secure the pushrod from the throttle spindle to the accelerator pump actuating arm. The diaphragm and spring can be removed from the housing and inspected for wear or damage. The well cover is secured by a further four screws and can be undone and removed from the carburettor body. There is a gasket between the cover and the body. The accelerator pump intake valve is screwed into the well cover housing. This is a one-way valve only.

The Weber carburettor has the accelerator pump intake and discharge valve fitted into the bottom of the float

DELLORTO DHLA - EXPLODED VIEW
1 Choke, 2 Auxiliary venturi, 3 Main jet, 4 Idle jet, 5 Pump jet, 6 Cold start jet, 7 Emulsion tube, 8 Cold start emulsion tube, 9 Air correction jet, 10 Idle jet housing, 11 Needle and seat, 12 Throttle butterfly, 13 Float, 14 Cold start tap screw, 15 Spring, 16 Cold start piston valve, 17 Set screw, 18 Cable clamp, 19 Float bowl cover, 20 Float bowl cover gasket, 21 Vent cover, 22 Vent cover screw, 23 Cover screw, 24 Spring washer, 25 Vent cover gasket, 26 Bolt, 27 Washer, 28 Cable nut, 29 Actuator cam, 30 Spring washer, 31 Sleeve, 32 Drive link, 33 Return spring, 34 Retaining washer, 35 Retaining clip, 36 Nut, 37 Filter, 38 Float fulcrum bar, 39 Washer, 40 Banjo (single), 41 Filter, 42 Washer, 43 Banjo bolt, 44 Banjo (double), 45 Washer, 46 O-ring, 47 Cap screw, 48 Pump jet spacer rod, 49 Pump jet metering check ball, 50 Pump jet holder, 51 Spring, 52 O-ring, 53 Washer, 54 Bypass screw, 55 O-ring, 56 Pressure tap screw, 57 Progression tap screw, 58 Needle valve, 59 Spring, 60 Washer, 61 Throttle (butterfly) shaft bearing, 62 Spacer, 63 Throttle balance screw, 64 throttle drive arm, 65 Throttle shaft locknut, 66 Nut, 67 Thrust washer, 68 Spring, 69 Adjustment nut, 70 Locknut, 71 Diaphragm spring, 72 Pump diaphragm, 73 Pump cover, 74 Washer, 75 Screw, 76 Pump housing gasket, 77 Pump housing, 78 Pump check valve, 79 Screw, 80 Throttle plate screw, 81 Throttle shaft, 82 Pump drive lever, 83 Drive lever screw, 84 Spring, 85 Stud, 86 AV set screw, 87 Spring, 88 Clip pin, 89 Compression pin, 90 Throttle drive arm, 91 Gasket and seal kit.
(Courtesy Dellorto).

© Dellorto

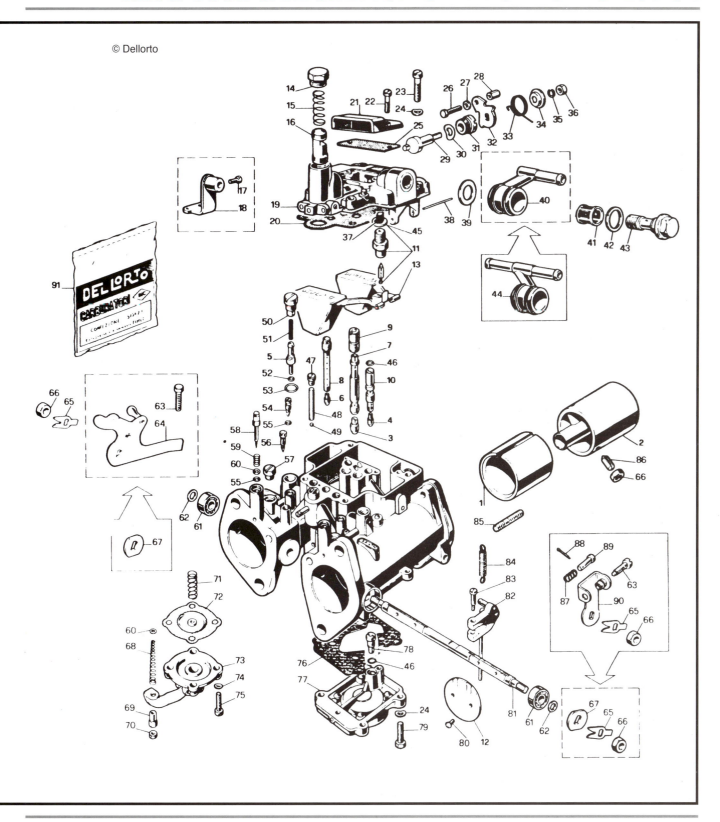

chamber and this can be removed with a large screwdriver. This valve can be quite tight and difficult to remove at times. The tapered-off top on this valve certainly reduces the size and strength of the screwdriver slot so it is important to use a screwdriver that fits the slot correctly in the first place. The Weber valve may have a small hole drilled in the side of it and this is part of the bleed-off for fuel during accelerator pump action. If there is a small hole in the side of the jet fuel escapes out of here back into the fuel bowl rather than being injected into the engine. This is part of the reason why Weber pump jets go up in 5s. The situation is alterable via the hole size found in the side of the jet. Valves are available that do not have a discharge hole. With the side hole blocked off or not present this accelerator pump intake jet is a one-way valve.

The two screw plugs above the accelerator pump check balls and weights can be removed using a correct fitting screwdriver and the carburettor turned upside-down so that the balls and weights fall out. (Use your large shallow tray for this purpose).

The Weber accelerator pump control rod is held in position by a brass retaining plate. To remove the accelerator pump control rod the retaining plate has two small indentations in it that can be used for the removal of the plate. This is carried out using long-nosed pliers that have had the ends chamfered (by grinding) so that the tips of the pliers go to the bottom of each indentation in the retaining plate so that, when the pliers are used to squeeze the plate, they have the maximum amount of contact with the sides of the indentations. Caution! Do not squeeze too hard or the plate will distort. The indentations' main function is to centralize the spring found underneath the retaining plate.

Stripping down for cleaning, checking and, if necessary, replacement of the basic working parts of the carburettor/s is now complete. (The fuel enrichment mechanisms can remain installed as they are not used; the Dellorto mechanism is excellent and never causes problems).

The carburettor body can now be thoroughly washed and all scale or corrosion carefully scraped off the inside of the body and all sediment removed by washing in a shallow bath of clean gasoline (petrol). Clean out the throttle bores and remove any corrosion or surface roughness with very fine wet and dry paper. If the cleaning fluid becomes dirty and appears to be carrying particles, change it. With the body washed, and clean, dry it off using compressed air (100 psi or more).

FUEL ENRICHMENT DEVICE - BLOCKING OFF DISCHARGE HOLES (WEBER ONLY)

The Weber carburettor can have the discharge holes blocked off to eliminate air/fuel mixture leakage into the main tract during normal operation if the fuel enrichment mechanism proves to be faulty (often a problem with well-used Webers), or even if you just wish to prevent future problems. This is done by tapping with a 6mm by 1mm pitch tap the outlet bore of each throttle bore for 12mm (0.5 inch) and installing suitable short grub screws into each one. The grub screws should be Allen-headed and must be securely fitted into the carburettor body. The tap used should be a taper tap or first tap which means that it will have a long lead in section. (This allows the thread of the grub screw to really wind into the thread cut into the body of the carburettor over a longer distance than if a plug tap was used). A firmly wound in grub screw coated with a sealer/

locking agent will never move. Check to see that no part of the grub screw protrudes into the throttle bore.

Weber also produce competition only carburettors which do not have a fuel enrichment device/choke mechanism. These carburettors have thin pressed steel plate to neatly block off the back of the carburettor. Starter valves are omitted and there are no holes drilled for them. There are no holes drilled into the throttle bore either, so there's no possibility of fuel leakage.

INSPECTING COMPONENTS FOR WEAR AND DAMAGE

Make the following checks as soon as the stripdown is complete so that you can order any parts which are not normally included in the basic repair kit. For each carburettor you'll need a new and full set of gaskets, O-rings and, advisably, diaphragms, so check what is included in the basic repair kit package. A new needle valve assembly is recommended.

Weber & Dellorto

Check to see if there is anything missing (see exploded view) from the carburettor/s you have dismantled. Check nuts for burring or damage. Check the fuel filter for damage. Check all screw heads for burring (clean them up or replace them). Ensure the floats are inspected for obvious damage and tested for leaks, their fulcrum pin checked for grooving or wear and the floats' hinge loops checked for looseness on the pin. Check needle and seat of the needle valve assembly for grooving around the point or tip (if grooved, replace both).

Check the throttle butterfly spindle for smooth and free movement (opening and shutting the butterfly). If there is any sign of 'grittiness' or 'lumpiness' remove the nuts and arms, covers or seals and check the condition of the bearings. If the bearings do not respond to the cleaning process suggested (see following text on difficult procedures), replace them with new items.

Every passageway must be checked and tested to ensure that it is clear of obstruction. Although the passageways change direction a lot, all can be shown to be clear by passing compressed air through them. If no compressed air can be felt assume until proven otherwise that the particular passageway is blocked. If a passageway has a blockage it must be cleared - see the following text on difficult procedures.

Weber only

Check the spindle return spring (they break frequently so it might be a good idea to fit a new one anyway). Check the accelerator pump control rod for end wear. Check the accelerator pump spring for breakage (rare).

Dellorto only

Check the diaphragm return spring for breakage (rare). Check the accelerator pump diaphragm for deterioration (very slow to deteriorate). Check the spindle return spring for breakage (rare). Check the cast aluminum accelerator pump actuation arm for cracks (occasionally happens). Check the fuel enrichment shut-off washer for deterioration (almost never wears out).

DIFFICULT PROCEDURES

There are four procedures that often pose difficulty: 1) cleaning all of the internal passages within the carburettor body; 2) removing and refitting the throttle butterfly, spindle and bearings; 3) removing broken threaded components from the carburettor body; 4) removing jammed chokes and auxiliary venturis.

Here are workarounds -

Clearing passageways

The carburettor body has many drilled and plugged passageways and - short of removing each plug and cleaning each passageway then replacing the plug - there are two ways of cleaning them. Note that it is not usually necessary to remove any plugs to clean the passageways. There are processes available today that can clean a carburettor inside and out to near new condition and most carburettor rebuilding companies use such processes. After having a carburettor body cleaned by a contractor, check each passageway to see that it is, in fact, clear using compressed air.

The alternative to professional cleaning is to wash the stripped down body in clean gasoline (petrol) and use compressed air (100 psi plus) with a large nozzle air gun to blow through all passages. Most carburettors respond well to this treatment; however, there is always going to be the odd carburettor with something firmly lodged in a passageway which will be detectable when blowing through each passageway with the compressed air as no air will be able to pass through. When testing for clear passageways always ensure that the nozzle of the air gun is well sealed off (tapered nozzle) against the hole so that all of the compressed air has to go through the passageway. The nozzle should have a 2.0mm (0.080in) hole in it so the passageways are subjected to a large volume of air at high pressure. If one passageway will not pass any air through it, that is the time to start removing plugs and checking the passageway physically for an obstruction. On odd occasions carburettors haven't worked correctly since new, or since being reconditioned, and investigation might

reveal that there is an obstruction in a passageway. Lead plugs have been found in passageways - it's rare but it happens. Always check the passageways to make sure that they are all clear on any carburettor which doesn't seem to be able to be tuned.

Throttle butterfly, spindle & bearings - maintenance, removal & refitting

Spindle bearing maintenance. It is very unusual for bearings and spindles to actually need replacing because of wear. Frequently the spindle bearings will feel quite 'gritty' but this seldom means that the bearings are worn out.

On Dellortos the solution to 'grittiness' in the bearings' action is to undo the spindle nuts and remove the throttle arm or cover which will expose the side of the bearing. Then, using a jeweler's small screwdriver, lever out the seal and wash the bearing out with gasoline. The seal will come out quite easily and without damage if care is taken. The gasoline will dissolve the dried grease and the action of the throttle will become smooth. Re-grease the cleaned bearings with a silicon grease or any other grease that is not affected by gasoline. Don't forget to refit the seals.

The Weber has bearings without seals, instead it has dust caps and spring covers and the spring cover is not always easy to remove. Each spring cover must be removed very carefully so as not to distort it (it's not available as a spare part). There are two small holes in the face and, ideally, a two-pronged pin wrench (spanner) is used to locate in these holes to facilitate a turning action. Suitably-sized snapring (circlip) pliers can be used to locate in the two holes and, using a small propane torch flame, apply localized heat to the bearing boss (carburettor body). With turning

A throttle spindle dust cap being removed from a Weber using circlip pliers (after the body has been heated).

force being applied to the spring cover bring the torch up to the carburettor body. Once the heat expands the aluminum of the body the spring cover becomes loose and can be lifted out as it is twisted. The heat required is not great (hot to the touch will not distort the carburettor body). The dust cover will lift out and will usually have to be replaced with a new item as they do deteriorate with time and use.

Wash the bearings out and if the 'lumpy' action disappears, re-grease the bearings and replace the dust cap and cover. Replace the bearings with new items if this cleaning process fails to improve the throttle action.

Butterfly, throttle spindle and bearings removal. The butterfly spindle bearings are not really highly loaded and bearing failure is usually caused by the sealing arrangement no longer keeping out moisture so that corrosion ruins the bearing. If the bearings have to be replaced, make sure that you have new butterfly screws, nuts and locking washers on hand as well as the parts to be replaced (spindle or bearings or both).

Make sure that the washer lock tabs are bent back flat before undoing the nuts on each end of the spindle. Usually the nuts can be undone quite easily using a six-point box end wrench (ring spanner). The spindle will have to be held while the nut is undone. Use long-nosed pliers (not overly tight, just firmly enough to hold the spindle) with aluminum strip between the jaws and the spindle to protect the spindle from damage. The advantage of using the long-nosed pliers is that the spindle can be firmly held (to preclude any twisting of the spindle) adjacent to the throttle bore wall. It seems a rough way of doing this job but it is effective and doesn't mark the spindle or butterflies.

There is no need to damage a steel spindle or the butterflies when removing the nuts at the end of the spindle. At the very worst the nut and washer are lost, but they are two of the cheapest items on the carburettor. If there is any doubt about the ability of the nut to come off easily, or if the nut fails to budge using normal force, move to the next procedure which is to cut the nut off. A brass spindle is easily damaged and will usually have to be replaced after it is removed as the threads will most likely have been damaged beyond repair.

With the carburettor clamped between bits of wood in a large vice, the throttle arm firmly hand-held, use a junior (very small) hacksaw to cut away a part of the nut adjacent to the spindle but not so close as to cut away the threads of the spindle. With the nut cut down to the washer on one side, move to the other side of the nut and cut that side down to the washer. The nut will now be exerting a minimum of clamping pressure on the throttle spindle and can almost certainly be wound off using the remaining flats of the nut.

The throttle butterflies are held in

Long-nosed pliers being used to hold the butterfly and spindle firmly while the nut on the end of the spindle is undone. There is aluminum shim between the pliers' jaws and the butterfly to prevent damage to the butterfly.

This nut has been cut in two places to release it from the spindle. Always cut off tight nuts to prevent spindle damage: nuts are cheap, spindles are not!

position by two brass screws and have the protruding portion of the screw threads crimped to stop the screw turning. The crimped portion of the screw threads can be removed (ground away) using a 4mm (3/16in) diameter ball-nosed rotary file spun at high rpm in a small hand grinder. This allows the removal of the screw even though the screw is effectively ruined. Without the crimping being removed the screws are virtually impossible to get out and the screwdriver slot is invariably damaged beyond further use. Once the screws are out the butterflies are slid out. Be sure to mark each butterfly for position so that it can be returned to exactly the same place (right way around, right way up) and the throttle bore from which it came. When fitting butterflies to the spindles use new screws and be sure to crimp them.

The accelerator pump arm must have its pin removed in the case of the Weber, and before the spindle can be removed from a Dellorto the screw which tensions the clamp around the spindle has to be undone, removed, and the clamp pushed over the flats and off the spindle altogether.

With the body held in a vice (between two pieces of wood) and with the spindle in the horizontal axis, the carburettor can be heated slightly around the spindle bearing housing with a small propane torch. Then the spindle can be tapped out, using very light hammer blows to the end of the thread from a copper hammer or a copper drift. With sufficient heat in the aluminum, tapping against the threaded portion of the spindle is not as bad an engineering practise as it sounds and the main resistance will, in fact, be the fit of the spindle into the bearing at the end which is being tapped. The other end will be reasonably free as the bearing will not have a full press fit with the body heated and the bearing must move with the spindle.

The spindles on Dellortos and Webers are stepped down where the bearings are fitted. The spindles are actually 8mm in diameter and step down to 7mm to fit the bearing bore's inside diameter. The threads at the end of the spindle are also 7mm diameter. The bearings are pushed into their respective bearing bores in the carburettor body until they bottom out. The length of the 8mm diameter section of the spindle matches this very closely, which gives the spindle minimal end float with near frictionless throttle action. Both carburettors are very accurately machined.

The reason for heating the carburettor slightly is to expand the aluminum. This means that for a short period of time the bore of the carburettor body has a less effective press fit on the bearing's outside diameter than when the relevant components are cold. Aluminum expands around twice the amount that steel does (coefficient of linear expansion) for the same given temperature rise. When the spindle is tapped out the bearing will come out of its housing easily and without any scoring of the internal surface of the housing bore. Whichever way the spindle is pushed out, one bearing must come out. That leaves one bearing still in the carburettor body and the other bearing fitted to the spindle.

If it is decided to replace the bearings on a Weber, the spring cover and spring and dust cover do not need to be removed separately. After the nuts and lockwashers, throttle arm, plain end cover, butterflies and accelerator pump control lever dowel have been removed, the spindle can be tapped out and these other components will automatically be pushed out. This method removes the possibility of damaging the spring cover.

The butterfly screws can usually be undone simply by turning them out, however, the screwdriver must be a perfect fit in the slot of the screw. Even though the screws are crimped, they'll usually wind out without damage, though it is always better to grind or file off the crimping before screw removal if you can. If you do re-use the screws, use thread locking compound and remove the protruding thread ends after installation because the previously crimped section will have become brittle and may break off and enter the engine.

The dowel that locks the accelerator pump control arm is removed with a pin punch that is a maximum of 1.9mm (0.074in) in diameter. This lever is usually reasonably loose on the spindle.

Next, the carburettor body should be heated gently to reduce the effective fit of the non-aluminum components and the spindle can then be tapped out. The spring cover will come out first, followed by the spring, the dust cover and then the bearing still attached to the spindle. The remaining

Throttle spindle bearing resting on top of vice jaws and the spindle about to be knocked out. Use a soft copper hammer or an aluminum drift to hit the end of the spindle. Tap the spindle square on.

Outer bearing race ground away enough to allow a specially made piece of flat bar to neatly fit into the bearing and be turned 90 degrees. A long drift can be used to push against the flat bar with the knowledge that it is securely locked into the race. Heat the carburettor body before shifting the race.

bearing is removed with a 8mm (5/16in) diameter rod with the spring cover, spring and dust cover coming out before the bearing.

The bearing fitted to the spindle can be removed by sitting the spindle loosely between vice jaws (1.5mm [1/16in] per side of the spindle) with the bearing sitting on top of the jaws and tapping the spindle out using a copper hammer or drift. The vice jaws must be flat across the top and a shim placed between the spindle and the jaws to protect the spindle.

To remove the last bearing polish down 8mm stock rod (or use a 5/16in diameter rod) 180mm (7in) long until it is clearanced to fit into the spindle bore and apply localized heating to the area around the bearing. The rod can then be inserted into the carburettor body until it contacts the inner part of the ball bearing and then the bearing can be tapped out using light hammer blows to the end of the rod - don't forget to heat the carburettor body.

If the bearing does not come out of the housing in one piece, it has totally collapsed and the outer race may now be firmly stuck in the housing of the carburettor body. There are a few ways that the bearing outer race can be removed. The first is to hold the carburettor body in a vice, between pieces of wood, with the offending bearing facing the ground and then, using a small propane torch flame, gently heat the bearing boss area of the carburettor body (from dead cold) and see if the bearing outer race simply falls out (aluminum expands and gravity causes the outer race to fall out).

The second way is to make a special tool out of 3mm deep by 10mm (0.125 by 0.375in) wide mild steel flat bar cut to a length of 16.5mm (0.65in). Radius the ends of the tool to give it, in effect, a 16.5mm (0.65in) diameter (to fit the groove in the bearing housing). Then, using a small high-speed grinder, grind away the edge of the bearing race in two places 180 degrees apart so that the strap can be fitted into the bearing and then turned thru 90 degrees to lock into place. It can then be used, in conjunction with a thin drift, to drive the bearing race out (after heating the carburettor body).

The third method of removing a stubborn outer race is to grind two grooves into the bearing outer (in line with the spindle axis) so that the bearing collapses and comes out in two halves. This requires care because it is very easy to grind too deep and cut into the aluminum of the body. When the grooves are nearly through the steel of the bearing there will be no strength left to hold the bearing outer in the hole as a press fit. This is the last resort method of removing the bearing outer and it will always work.

The bearings should be a light press fit in the carburettor body (size for size) but invariably the odd bearing is going to be loose in the casting. Check to see if the new bearings are going to be a light press fit in the carburettor body. If the new bearings are loose and simply fall in, a proprietary retaining compound can be used to correct the situation. The fit tolerance between the bearing being a light press fit or just falling into the bearing seat in the carburettor body is only a matter of a thousandth of an inch or hundredths of a millimeter.

Note that spindles do get twisted from time to time which, ultimately, means that one butterfly per carburettor does not close fully. This only happens when the spindles/ linkage has been subjected to extreme axial twisting pressure at full throttle when there is no effective throttle stop. On secondhand carburettors it's always worth checking to see that both the butterflies actually close. A symptom of a twisted throttle spindle is high idle speed when the throttle adjustment screw is fully backed off. The usual solution is to replace the spindle/ spindles, although some people do straighten them by twisting them back.

Spindle and bearing refitting

Assuming that all bearings have been removed from the carburettor body, proceed as follows.

On the Dellorto the spindle goes in one way only. The accelerator pump arm slot is positioned on the right hand side when the carburettor is viewed from the butterfly side.

The Weber accelerator pump control arm fits on one way only and the spindle has to pass through it now but the spindle can go in either way. The lever faces inwards and must be the right way up (check the exploded diagram). If the accelerator pump control lever is new, check to see if it goes over the spindle. If it doesn't it will have to be reamed out with an adjustable reamer until it does.

Next, the spindle is fitted with a bearing (at either end). Put some oil on the end of the spindle and in the bore of the bearing. Place the bearing over the end of the spindle and you will note that a portion of the thread will protrude (about 3.3mm/0.132in). This means that the spindle is located into the bearing's bore. You will also note that the threaded portion has a flat on it. Using a good vice which has a level top, open the jaws up to 5mm (0.20in) and place the protruding threaded end of the spindle into the gap between these jaws. With the bearing on the top of the vice jaws and the spindle in the vertical position tap the spindle into the bearing. The jaws are opened wider than the spindle size to allow the spindle to pass on down without interference. Tap very gently (square on to the spindle) using a copper mallet. The spindle will go in slowly and, when it can go no further, the sound of the tap will change (heavier sound) as the spindle bottoms out on the bearing inner race.

The spindle is reasonably strong and it is acceptable to fit the bearing

 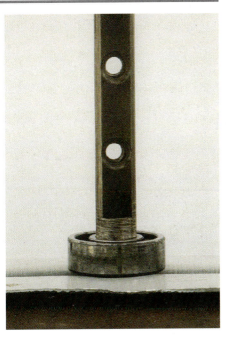

Photos show a spindle being tapped into a new bearing. Use a copper hammer or an aluminum drift to tap the spindle into the bearing. The spindle is guided to a large degree, but do make sure the spindle is square to the bearing and that the bearing is placed flat on the top of the vice jaws. Lubricate the bearing bore and spindle.

A new bearing in position on the spindle (which is there to act as a guide while the bearing is gently tapped into the carburettor body). The spindle just goes back as the bearing goes into the bore of the carburettor body.

on to the spindle in this manner. The   bearing and spindle are, more or less,

A pair of long-nosed pliers holding butterfly/spindle (with alloy shim between the pliers' jaws) while the spindle nut is tightened.

self-aligned before fitting because of the thread diameter (being 7mm) acts as a pilot.

With the carburettor held in a vice between two pieces of wood, insert the spindle into the carburettor body (the correct way around) and push it all the way in until the bearing contacts the side of the carburettor. Leave the spindle in this position and place the other new bearing to be fitted on to the threaded portion of the other end of the spindle. Run some oil around the outer diameter of the bearing. Apply heat to the bearing area of the carburettor body and, using a long socket 19mm (3/4in) in diameter and a small hammer, tap the bearing into the carburettor bore. As the bearing enters the carburettor the spindle will move

out of the other end of the carburettor.

The spindle is fitted in this manner to act as a guide for the bearing. By being fitted to the spindle the bearing

A spindle, with the bearing already pressed on, being tapped into the carburetor body. The other bearing has already been tapped into place. To prevent that bearing from moving, a washer is placed between the bearing and the flat surface of the vice top. This method requires assistance.

is more or less dead in line with the spindle and bearing bore axis. There is little chance of 'picking up' the aluminum of the carburettor body when it is fitted.

It is standard engineering practise to use heat when fitting steel parts into aluminum items that are supposed to be a press fit when assembled. This is done to ease fitting and make sure than the aluminum surfaces in contact are not scored during assembly and that the size of the aluminum part is always maintained. This is not to say that the spindle and bearings cannot be installed satisfactorily with all parts cold; they can be and usually are.

An alternative way of fitting the bearing (without heating) is to oil the outer surface of the bearing and the inside of the carburettor boss and, using a squared-off piece of tube or a long socket the same diameter as the bearing outside diameter, tap the bearing into the bearing housing. Not too much force is required to do this. During this process continually check to see that the bearing is square on to the bearing housing so as to avoid misalignment and possible damage to the carburettor body.

The next operation is to press the spindle into the centre of the bearing already installed in the carburettor housing and, at the same time, the bearing already fitted on the spindle is also fitted into the carburettor housing. Oil the spindle shank to be pressed into the bearing and oil the inside of bearing bore in the carburettor body. Oil the outer surface of the bearing already fitted to the spindle and oil the aluminum of the bearing bore of the carburettor body.

The carburettor body is held manually (get assistance) in the vertical position on top of vice jaws with a 3mm (1/8in) thick washer 19mm (3/4in) diameter and a hole not less than 7mm (5/16in) in diameter or more than 9mm (3/8in) in diameter between the bearing and the top of the jaws. This is to prevent the bearing being pushed back or out of the carburettor housing as the bearing already fitted to the body is actually recessed slightly. With the washer in place and the end of the spindle between the vice jaws set at not less than 5mm (0.20in), apply heat to the area around the bearing bore in the carburettor. Using a long socket with a nominal inside diameter of 7mm (0.312in) and a depth of at least 10mm (0.400in), tap the bearing into the carburettor body. With the body heated, the main resistance is going to be the pressing of the spindle into the bore of the bearing on top of the vice jaws. The bearing and spindle are seated when the sound of the tapping becomes more 'solid' as the spindle contacts the side of the bottom bearing and the top bearing contacts the aluminum of the carburettor body. Allow the carburettor body to cool.

On Weber carburettors make sure that the bearings are packed with grease then fit the dust cover and spring and finally the spring cover. Oil the edge of the spring cover and tap it into place using a long socket with an outside diameter of 19mm (3/4in). Tap the spring cover into the body until it is below the height of the shoulder where the 7mm diameter of the spindle starts: this way it will be out of the way of the throttle lever or end washer and will not interfere with the opening of the throttle. Fit the throttle arm to its end of the spindle and tighten the nut reasonably tight so it is easy to open and shut the throttle while checking the butterfly positioning. Make sure that the slot in the throttle arm is a tight fit on the spindle: if it isn't, the throttle arm will have to be replaced with a new item.

The Dellorto has both side bearings sealed and has a small spacer fitted between the bearing and the end cover or throttle arm. This is a much less fussy system than on the Weber.

Both carburettors have lockwashers and nuts to be tensioned and tabs to turn up to lock the nuts in place. The washers are slotted and locate on to the spindle flats. The nuts can only be tensioned while the spindle is firmly held otherwise the spindle can be damaged. Long-nosed pliers with aluminum shim (0.1mm/0.040in) between the jaws and spindle is quite satisfactory. The pliers' jaws should be positioned in the throttle bore adjacent to the nut being tightened and next to the throttle bore wall. That completes the spindle and bearing fitting.

The accelerator pump mechanism can now be fitted. With the Dellorto the arm slides over the spindle (full throttle position) and the screw is inserted into the hole and tightened and the top spring connected. The Weber uses a pin to locate the accelerator pump lever arm on the spindle and to hold it in place. The pin is tapped into the arm until it is flush.

Fitting throttle butterflies. Dellorto butterflies are in the correct way around when you can see the numbers below the spindle and the progression holes sweep recesses (if present) at the top of the throttle bore. Weber numbers are positioned in the same place as Dellorto and sometimes on the other side of the butterfly, which means that the numbers are inside and out of view. Some butterflies have sweep slots and some do not. If the butterflies are not centralized correctly the engine may never idle as slowly as it should because the throttle can't be fully closed (shut tight to eliminate all air passing through).

Fit one butterfly through the slot

in the spindle and put the two screws in but do not tighten them yet. Open and shut the throttle a few times to get the butterfly situated as centrally as possible and then tighten the two screws. Now look in through the throttle bore from the trumpet side and see how much light there is around the outer diameter of the butterfly. The amount of light must be equally distributed around the periphery of the butterfly and be as thin a band of light as possible (0.025mm/0.001in or less). If the butterfly is offset, more light will be visible on one side than on the other. There is a certain amount of clearance between the holes in the butterfly and the screws to facilitate adjustment of the butterfly's position to optimum.

Once one butterfly is set correctly move on to the other butterfly and if there is a problem in getting that butterfly central at least you know which butterfly is causing the problem. If the two butterflies are being set up together it can be difficult to actually see which butterfly is not central. It is quite possible to get the butterflies to shut off completely.

If a butterfly cannot be centralized this will usually be due to an unfortunate combination of manufacturing tolerances. The holes in the butterflies will have to opened out (elongated) using needle files to give more clearance to the screws. To check for which way to remove material from the butterfly remove the screws and shut the throttle carefully. It does not take much butterfly deviation from the central position to cause the throttle to jam. Depending on the amount of mismatch it will usually be possible to see a part of the butterfly through the holes in the spindle. Remove that part of the butterfly that you can see. If it proves difficult to see the butterfly, remove it and polish it so it is easier

Dellorto butterflies with the progression hole sweep slots and the numbers clearly visible.

to see. Hand file the butterfly hole for more clearance until the butterfly fits properly.

The fit is correct when both butterflies have as near as possible equal bands of light around their respective diameters (view this from the trumpet side of the throttle bore). **Note**: make sure that the throttle arm stop adjusting screw is wound back well out of the way so that it is not interfering with the shutting off of the throttle during the butterfly fitting and checking procedure.

Remove each screw in turn (only one at a time) and apply some proprietary thread locking compound to the thread. Refit the screw and tension it as tightly as possible within the confines of the strength of the screwdriver slot. **Caution!** The protruding screw threads should be crimped as well just to be absolutely sure that the screws don't come out and end up going into the engine. The spindle (screw heads actually) has to be supported or rested on a bar of aluminum which goes up

into the throttle bore. An aluminum flat bar 35mm by 10mm (1 5in by 0.375in) clamped in a vice with about 50mm (2in) sticking above the jaws will provide a suitable rest for the spindle. With an assistant to hold the carburettor in position, use a long pin punch to get down the carburettor throttle bore from the trumpet side and peen over the top of each side of each screw sufficiently so that, even if a screw did come loose, there is no way it could wind out of the spindle.

Check to see that the butterflies are at 90 degrees when the throttle is fully opened. If they are over centre the stop on the throttle arm is not in the correct position. Fit a new one or put a run of braise on the stop and file it down until the butterflies are 90 degrees with the lever on the stop.

Removing damaged threaded components

When working on these carburettors it is important to use the right screwdriver head size for the particular slot size of the screw plug and a six-

sided box end wrench (ring spanner) and not a twelve-point to avoid damaging components. Reasonably well-maintained carburettors handled with reasonable care never get into a poor state of repair.

Threaded jets and other threaded components can usually be easily removed. Some components, however, such as the idle adjustment screws found in the towers of some Dellorto carburettors, become well and truly jammed, or the screwdriver slot may no longer be usable. This is quite a common problem with those fine thread Dellorto idle adjustment screws in towers. The carburettor is useless in this condition, in terms of getting the idle mixture of that particular cylinder correctly set. The only solution is to remove the damaged screw.

The carburettor body will then have to be taken to a precision engineer to have the damaged part bored out. To do this, the carburettor body is mounted in a machine vice which is bolted to the table of a vertical milling machine (Bridgeport or similar), and the centre of the damaged part lined up perfectly with the spindle.

Dellorto carburettors with towers have very fine threaded idle mixture adjustment screws which frequently seize solid and the screwdriver slot gets damaged.

Photos show a simple-to-make puller for the removal of jammed chokes.

A choke puller installed. The puller's bottom plate is behind the choke to be removed and the nut is about to be tensioned. The carburettor body may have to be heated to get the choke to move.

The centre of the jet is then milled out with a milling cutter and a drill used to take out the final amount of material so that all of the component is removed without damage to the threaded portion of the carburettor body. The milling cutter (three flutes) or a slot drill (two flutes) is used in the first instance as it will bore true. Twist drills tend to wander in anything less than perfect circumstances (these are not perfect circumstances) and are used only to open out an existing true bored. The remaining thread of the damaged idle adjustment screw is 'picked' out with an engineers scriber.

Removing jammed chokes and auxiliary venturis

Usually the chokes and auxiliary venturis will come out easily. If they do not it is usually because of corrosion and the ensuing build-up which effectively removes the sliding fit of the choke and auxiliary venturi in the throttle bore. Chokes, and to a much lesser extent auxiliary venturis, can become severely jammed and a considerable amount of force can be required to remove them. Virtually all jammed chokes can be removed without damage to either the carburettor body or the choke.

Check to see that the locking screw has in fact been removed from the carburettor body. Auxiliary venturis are removed by inserting a length of 12.7mm (0.5in) wood dowel through the back of the carburettor throttle bore (one side of the opened butterfly) and contacting the bridge section edge that connects the outer diameter of the auxiliary venturi to the central nozzle. The Weber has two sections that join the outer diameter of the auxiliary venturi to the nozzle, while the Dellorto has one. The wood dowel can be alternated between the two sections when removing Weber auxiliary venturis, while the Dellorto only has one section.

It is not usual for the auxiliary venturi to be too tight. Spray penetrating oil down the throttle bore from the trumpet side of the carburettor and wait for ten minutes for it to soak in before continuing. The wood of the dowel will become indented and possibly split but there will be no damage to the auxiliary venturi. Light hammer blows to the end of the dowel are all that is required to remove the auxiliary venturi.

The chokes on the other hand, can be well and truly tight but once again they invariably come out. The chokes are drawn out using a custom-made puller as shown in the accompanying photos.

The plate that fits in behind the choke has to be made to suit the actual choke which is to be removed. The end plate is made 2mm (0.093in) larger than the diameter of the choke to be removed. The end plate is made out of 20mm (0.78in) by 6mm (0.236in) mild steel bar and has a hole drilled through the centre. This hole is then elongated using a file so that it becomes 11mm (0.437in) long. The reason for the elongated hole or slot is to allow the end plate to be angled so that it can be inserted in through the choke. The outer edges of the end plate are chamfered to suit the taper of the choke so that both tapers fit each other as closely as possible. As a precaution against marking the inside of the choke with the bottom plate, place some thin cardboard between the end plate and the choke's surfaces before the bottom plate is pulled up to contact the bore of the choke.

The main puller rod is 8mm (0.312in) diameter threaded bar stock and is cut to 110mm (4.25in) long. One end of the puller has a Nylock nut (a half nut or a cut down standard Nylock) and the other is just a plain hexagon nut.

The other parts of the puller are the same no matter what carburettor type or size is having a choke removed. The outer end plate is made of 20mm (0.78in) wide by 8mm (0.314in) thick mild steel flat bar and will have to be 55mm (2.16in) long and have a hole drilled through the middle with clearance on it to suit the diameter of the main puller rod.

The puller is inserted into the carburettor's throttle bore with the plate angled (as shown in the photo) and through the actual choke. The angled position of the plate allows the bottom plate to pass through the choke diameter and, when straightened, lock into the taper of the choke.

The top nut is wound down firmly on to the top plate (by hand) and then tensioned with a wrench (spanner), which will ultimately remove the choke. If the choke still refuses to move even with considerable tension on the nut, keep the tension on and heat the carburettor body (approximately where the choke is) with a small propane gas torch. The body of the carburettor will absorb the heat and expand and, in so doing, momentarily become larger than the choke's outer diameter before the heat is transferred to the

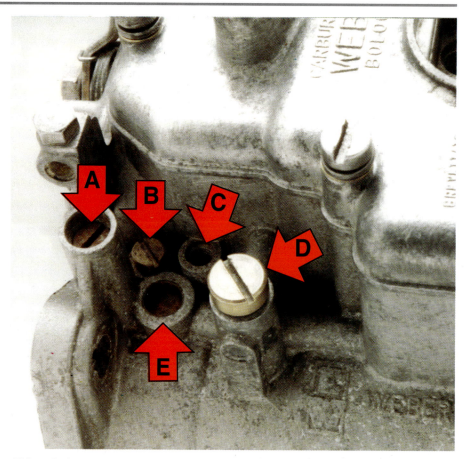

This emission controlled 40mm Weber carburettor has the idle mixture adjustment screw housed in a tower (A), an 'idle by-pass' screw plug (B), recessed 'idle by-pass' adjustment screw (C), accelerator pump top (D), and the progression holes' permanently pressed in brass plug (E).

choke. Minor hot-to-the-touch heat only is applied to the carburettor body (excessive heat may distort it).

With the choke removed, clean the throttle bore using 280 or 320 grit wet and dry paper and kerosene (paraffin) until the surface inside is as near to polished as possible. The surface will not usually take a polish because it will be too stained. Clean the choke's and the auxiliary venturis' outside diameter with wet and dry abrasive paper and the choke and auxiliary venturi will then be an easy fit into the throttle bores (the sliding fit or tolerance is approximately 0.1mm/0.004in).

Recognising 40mm emission controlled sidedraught Weber and Dellorto carburettors

It's only the 40mm sidedraught Weber and Dellorto carburettors that have been subjected to the emission control modifications. It's quite vital that you can recognise these Weber and Dellorto carburettors because, while they certainly look the same as other models, they aren't, and they can be less suitable for racing applications than the universal performance carburettors of either manufacturer. You have to know what to look for, and only buy carburettors that are going to be suitable for your intended application.

This late model emission controlled 40mm Weber carburettor has a different type of accelerator pump. It's externally operated and quite similar in principle to the Dellorto carburettor.

The Dellortos and the Webers have slightly different internal circuitry to achieve what is ultimately the same thing - that is, make the idle jets take fuel off the main jet well. The Dellorto has the additional feature of having a fixed idle jet air bleed at 2.0mm in diameter. The Weber doesn't have this and maintains the standard system. Emission Dellortos also have an identifying stamp on the body where the lettering DHLA40 is. The C and E model Dellortos are universal performance non-emission models, while the F, G, H, L and N models are emission carburettors. There are small differences between the several versions of emission control carburettors made by Weber. Some versions, for example, have external accelerator pumps, though most do not. Some have very small front holes for float bowl venting, and some have

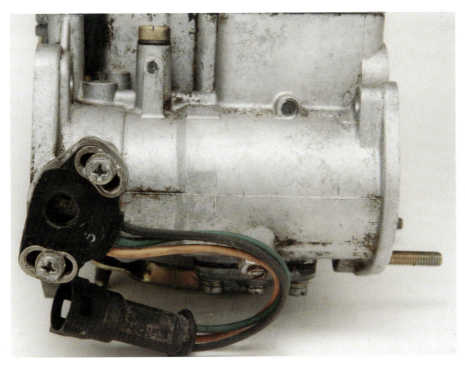

This late model emission controlled 40mm Weber carburettor has a potentiometer fitted for throttle position sensing.

Dellorto emission control carburettor has high mounted air bleeds to the idle jets. Take the cover off to check the position. Look for the brass sleeves clearly visible in the two small recesses as arrowed.

This Weber emission carburettor body has 'O' ring type idle jet holder as shown at A and B, small float chamber vent hole at C and tower type idle jet adjustment screw towers at D.

Accelerator pump inlet valve is on the underneath side of this Weber carburettor body adjacent to the main jet inlet holes. These carburettors also have an shut off valve in the accelerator pump circuitry.

vented inspection covers. What both companies' emission type carburettors have which seriously affects the progression phase for acceleration purposes is a lot of progression holes (five large holes compared to three small holes, for example). This is a real giveaway and allows the carburettor to be categorised from this factor alone. Lots of progression holes (five) of large diameter means a lean progression phase.

Dellorto emission controlled carburettors recessed idle by-pass adjustment screw. Note 40H visible lower left with the H indicating its emission status.

The three small progression holes in this 'universal performance' type Dellorto carburettor is what you would expect to find on a non-emission type Dellorto. Some Weber carburettors, for example, have only two small diameter holes.

Chapter 2
Rebuilding

With all parts spotlessly clean and an equally clean working area, the rebuild can begin. Check the position of all parts against the exploded view of your particular carburettor/s. The rebuild procedure that follows also lists the parts that need to be checked for wear and possible replacement.

Dellorto needle and seat on the left (note the small filter on the top of the seat), Weber needle and seat on the right.

NEEDLE VALVE & SEAT (DELLORTO)

Assembly starts with the fitting of the needle and seat body into the carburettor top. Put the aluminum washer over the seat body and screw it in. Using a six-sided box end wrench (ring spanner) tighten using reasonable tension. There is no need to go too tight as the thread is not large. Place the gasket on to the carburettor top as this must be in position before the floats are installed.

FLOATS & FULCRUM PIN (DELLORTO)

Fulcrum pin - checking

Check the floats' fulcrum pin for wear grooves and, if there is any wear, replace it. The pin, after some use, will have rub marks on its surface but this does not mean it is actually worn. Unacceptable wear can be measured with a vernier calliper or micrometer. The pin first goes into the hole in the post without a slit as this is a clearance hole for the pin. The post with the slit is the one that applies tension to the pin by lightly clamping it when the pin is pressed into position. The post is able to expand slightly when the pin is pressed in because the slit allows movement.

Floats - checking

If the carburettor is known to have done a lot of work, replace the floats. If they appear to be in serviceable condition place the pin in through the eyes and check to see how much movement there is between the pin and the eyes. Shake the float and listen for the sound of fuel sloshing around inside it as this indicates a leak. Place the float in a bowl of fuel to check that it floats. Immerse the float in the fuel and check for air bubbles. Air bubbles indicate an air leak. Check new floats for leaks in the same manner, just to be sure.

The pin should be able to rotate quite freely but any excessive slackness should be removed. This is done by squeezing the floats' hinge loops using needle-nosed pliers. The aluminum can be squeezed in a progressive manner to fit the pin very closely but not too closely as this may cause the float to bind on the pin during operation, which could lead to flooding. The pin fit in some floats is very loose which is not conducive to accurate fuel metering by the floats.

Floats & pin - fitting

The needle is fitted into the seat body first, then the floats assembly is lined up with the posts and the spring-loaded head of the needle and moved across so that the two small tabs fit under the head of the needle and seat. Fit the fulcrum pin into the post (without the split in it) and push it through to the second post (the one with the split). When the pin is close to the split post, line it up exactly with the hole and carefully tap it in. The pin should be tight when fitted. If the pin goes in very easily the post may not be putting any tension on the pin. If so, place the pin between the posts so that the ends of the pin protrude an equal amount from each post and, using pliers, squeeze the split post slightly. It doesn't take much to 'tighten' the hole so that it exerts sufficient clamp to retain the pin.

FLOAT LEVEL - SETTING (DELLORTO)

There are three floats for DHLA carburettors, numbered 7298.1, 7298.2 and 7298.3. Over 95% of DHLAs use the 7298.1 floats (which are 8.5 grams) and have the 15mm/0.590 inch fuel shut off height. The 7298.2 has a 17mm/0.670 inch fuel shut off height and is 8.5 grams in weight. The 7298.3 has the same shut off height as the 7298.1 but is 7.0 grams in weight. Contact a main agent

Dellorto floats being set at 15mm when the needle is just seating.

Dellorto floats being set to 25mm droop.

to find out what your float/s should be if you are in any doubt. Quote the model number cast and stamped on the lefthand side of the body.

The float height of the vast majority of floats is 15mm (0.594in) and it is measured with the gasket in place. A rule can be used to measure from the gasket to the top edge of the float. The needles are spring-loaded which means that the needle will seal on its seat but the float will still keep moving. The float level is measured when the needle just contacts the seat. This measurement is checked with the carburettor top held on its side. If the top is tilted the needle will be seen to move out from the seated position. If the top is slowly tilted back the needle will move in towards the seat and, when it stops moving, this is the point at which the distance between the float and the gasket must be measured.

The two floats must be set equally. This is achieved by twisting the aluminum arms using long-nosed pliers. The arms bend reasonably easily and it is possible to set the floats exactly as they should be set. The droop setting is 25mm (0.985in) and this is controlled by the tab at the back of the float and the float stops dropping when this tab contacts the seat body. This tab can be bent with long-nosed pliers to reduce or increase the droop measurement.

TOP COVER (DELLORTO)

The carburettor body is prepared next so that the top can be positioned and screwed on to it. The starter jet is screwed in (if it was removed), the accelerator pump check balls and then the weights are placed in their respective holes. The screw-in plugs are fitted above these two items.

The top is positioned on the carburettor body, and the four screws inserted into their respective holes and

The jetting that has to be in place before the top cover is fitted.

Carburettor top cover fitted and four securing screws (arrowed) tightened.

tightened progressively. One screw will have a tab under its head; this is the reference number for the particular carburettor. **Caution**! The floats are less vulnerable to damage if the cover carrying them is installed as soon as possible after setting the float height.

Caution! Make sure that all of the screws used to secure the carburettor parts to the body have washers fitted between their heads and the body. This will prevent any damage to the aluminum of the castings.

Dellorto carburettor top with the inspection cover off showing the emulsion tube holders and the idle jet holders.

Choke and auxiliary venturi positioned in front of a Dellorto carburettor. Fit the choke first, then the auxiliary venturi.

Dellorto externally mounted idle mixture adjustment screw components (above) and correctly assembled (below).

BODY COMPONENTS (DELLORTO)

The main jet, emulsion tube and air corrector holder combination are assembled and inserted into their respective holes and tightened.

The idle jet and idle jet holder are assembled and inserted into their

Dellorto idle mixture adjusting screws (arrowed).

Dellorto progression hole cover plugs (arrowed).

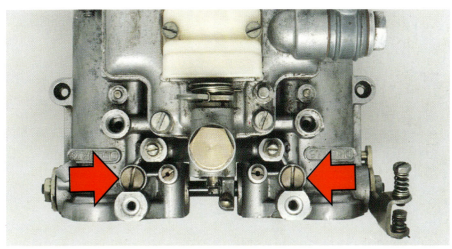

Dellorto accelerator pump jet and cover plug individual components (top) and assembled (bottom).

Dellorto accelerator pump cover plugs (arrowed).

Dellorto integral fuel filter and fuel union components.

respective holes, tightened and then the cover is placed over them and its two screws tightened.

The chokes first and then the auxiliary venturis are now inserted into the carburettor body and the retaining screw turned in until it firmly contacts the auxiliary venturi. The nut is wound down till it contacts the carburettor body and tensioned using a six-point

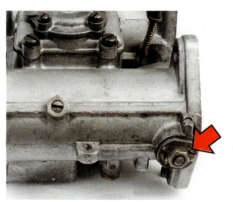

Dellorto with plain throttle spindle cover (arrowed) correctly fitted. Note how tab of lockwasher is turned up to lock the nut.

Choke location varies on some models of Dellorto carburettors. This 48mm DHLA uses screw location. Check to see that the hole in the choke is correctly lined up before winding the screw in as, otherwise, the choke can get damaged!

Dellorto. It's very important that the throttle arm is a tight fit on the throttle spindle.

Dellorto throttle arm in position on the throttle spindle. Note raised tab to lock nut.

Dellorto accelerator pump bottom well cover fitted to actuating rod.

box end wrench (ring spanner). Reasonable tension is applied to the nut while the screw is held with a screwdriver. This method of screw and nut locking only comes loose if the nut is not secured correctly.

Screw in the idle mixture adjusting screws. Make sure that the spring, washer and the small rubber O-rings are positioned in this order before screwing them in.

Screw in the progression hole plugs.

The accelerator pump jets are pressed into the screw plug and then inserted into their respective holes and tightened. Make sure that the small

rubber O-ring is fitted to the pump jet and that the fibre washer is fitted under the screw-in plug.

The fuel union is fitted next. If the fuel filter screen is damaged replace it with a new item. Fit new washers each side of the fuel union (sometimes called a 'banjo'). The large holed fibre washer goes against the carburettor body while the smaller holed one goes against the union bolt. Do not tighten this bolt at this point. Wait until the actual fitted position is known and always use a six-point box end wrench (ring spanner).

Fit the small washer and then a cover to the end of the spindle and fit a

Dellorto accelerator pump bottom well cover fitted to carburet tor.

new lock washer before putting the nut on. The nut must be tensioned with the spindle firmly held. This is done using long-nosed pliers with some aluminum shim between the jaws and spindle to protect the spindle from damage.

The throttle arm is fitted next. The first thing to check is the fit of the slot of the throttle arm with the flat of the spindle (the round, plain end covers are not critical here). If there is any looseness in fit the throttle arm must either be replaced or repaired. It should be a tight fit on to the spindle. Fit a new washer and tighten the nut. Use long-nosed pliers (in conjunction with some shim to protect the butterfly) to hold the butterfly/spindle while tightening the nut. Now check to see that when the throttle is fully opened and on the throttle stop, the butterflies are 90 degrees to the throttle bore of the carburettor. If not, the throttle arm will have to be altered (brazed) or an alternative throttle arm fitted that allows the butterflies to to attain the correct angle. When you are sure that everything is right, bend the tabs over on both nuts to secure the nuts.

The bottom well cover is fitted next but, before it is, fit the accelerator pump intake valve or one-way valve. The ball inside the valve must be free to move. With the carburettor turned over place a new gasket on the body and fit the well cover, then fit and tighten the four screws.

The accelerator pump mechanism is fitted next. The actuating arm is connected to the diaphragm housing by a rod with a spring over it which in turn is connected to the diaphragm actuating arm. Before the diaphragm housing is screwed on to the bottom of the carburettor the arm must be connected to the spindle. To do this the throttle is opened fully and the arm pushed over it. When the throttle is closed the arm goes in behind the

Weber needle valve and seat fitted into the top cover.

spindle. The spring is fitted into the carburettor body and the diaphragm into the diaphragm housing. The housing is then fitted to the carburettor body and secured with the four screws. Fit the spindle return spring.

At this point the Dellorto carburettor is assembled ready to be fitted to the intake manifold.

NEEDLE VALVE & SEAT (WEBER)

The rebuild starts with the fitting of the needle and seat body into the carburettor top. Put the aluminum washer over the seat body and screw it in. Using a six-sided box end wrench (ring spanner) tighten it using reasonable tension. There is no need to go too tight as the thread is not large. Place the gasket on to the carburettor top as this must be in position before the floats are installed.

FLOATS & FULCRUM PIN (WEBER)

Fulcrum pin - checking

Check the floats fulcrum pin for wear grooves and, if there is any wear, replace it. The pin, after some use, will have rub marks on its surface but

this does not mean it is actually worn. Unacceptable wear can be measured with a vernier calliper or micrometer. The pin goes first into the hole in the post that has no slit in it as this is a clearance hole for the pin. The post with the slit is the one that applies tension to the pin by effectively lightly clamping it when the pin is pressed into position. The post is able to expand slightly when the pin is pressed in because the slit allows movement.

Floats - checking

If the carburettor is known to have done a lot of work replace the floats with new items. If the floats appear to be in serviceable condition place the pin in through the float hinge loops and check to see how much movement there is between the pin and the floats. Shake the float and listen for the sound of fuel sloshing around inside as this indicates a leak. Place the float in a bowl of fuel to check that it floats. Immerse the float in the fuel and check for air bubbles. Air bubbles indicate an air leak. Check new floats for leaks in the same manner, just to be sure. Because Weber floats are made of brass they can, if the repair area is small, be soldered.

The fulcrum pin should be able to rotate quite freely in the float hinge loops but any excessive slackness should be removed. This is done by squeezing the floats' hinge loops using needle-nosed pliers. The brass can be gently squeezed in a progressive manner to fit the pin very closely but not too closely as this may cause the float to bind on the pin during operation which could lead to flooding. The pin fit in some floats is very loose and this is not conducive to accurate fuel metering by the floats.

Floats and fulcrum pin - fitting

The needle is fitted into the valve seat

body first and the float is lined up with the posts and the spring-loaded head of the needle and moved across so that the two small tabs fit under the head of the needle and seat. Fit the fulcrum pin into the post (without the split in it) and push it through to the second post (the one with the split). When the pin is close to the split post, line it up exactly with the hole and carefully tap it in. The pin should be tight when fitted. If the pin goes in very easily the post may not be putting any tension on the pin. If so, place the pin between the posts so that the ends of the pin protrude an equal amount out from each post and, using pliers, squeeze the split post slightly. It doesn't take much to 'tighten' the hole so that it exerts sufficient clamp to retain the pin.

FLOAT LEVEL - SETTING (WEBER)

The float shut off height is set to 0.295in/7.5mm. Weber list several float shut off heights for different models of carburettor, but consider 0.295in/7.5mm to be average and always usable. The height is measured with the gasket in place. A rule or vernier calliper can be used to measure from the gasket to the top edge of the float. The needles have a spring-loaded ball in the end which means that the needle will seal its seat but the float will still keep moving. The float level is measured when the needle just contacts the seat. This measurement is checked with the carburettor top held on its side. If the top is tilted the needle will be seen to move out from the seated position. If the top is slowly tilted back the needle will move in towards the seat and when it stops moving this is the point at which the distance between the float and the gasket must be measured.

The two floats must be set equally. This is achieved by twisting the brass

Weber floats being set to 7.5mm with the needle and seat just touching.

arms using long-nosed pliers. The arms do not bend very easily but, with a little patience, it is possible to set both floats exactly as they should be set. The droop setting is 15mm (0.694in) and this is controlled by the tab at the back of the float: the float stops dropping when this tab contacts the seat body. This tab can be bent with long-nosed pliers to reduce or increase the droop measurement.

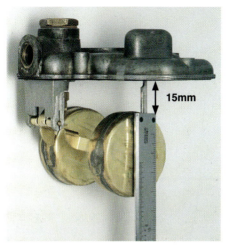

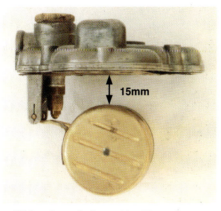

Weber floats being set to 15mm droop.

Late model (mid 1980s-on) Spanish built Webers with plastic floats

Although Weber used brass floats for about 25 years (mid -1980s), they changed to the current style of float. All of the current Spanish built sidedraught Webers which came out in the early 1990s have black 'Spansil' floats. 'Spansil' is a trade name for butadiene-acrylonitrile copolymer. They are excellent and have different float level settings to the earlier brass ones. The float level heights are measured with either a metric rule or a vernier caliper, and the float lever arm must be just contacting the needle and seat.

Weber also introduced a black neoprene material carburettor body to

Weber accelerator pump control rod height being measured.

A selection of assembled Weber accelerator pump rods, springs and plungers. The springs vary in tension and the pump rods vary in length.

A selection of accelerator pump rods without springs and plungers.

DCOE 151 and 45mm DCOE 152 versions with both having 'idle by-pass' circuitry and cast in towers on each side of the body to accommodate this system. They also have fuel enrichment mechanisms for cold starting purposes. The Promek made Spanish built carburettors are also available in 48mm, 50mm and 55mm sizes as DCO/SP versions, with the SP standing for 'special' (racing purposes). None of these three DCO carburettors have 'idle by-pass' circuitry, and top cover gasket. This material is less rigid, slightly thicker, and more able to guarantee a perfect seal.

The later Promek made Spanish built sidedraught Weber carburettors (Promek took over from Magneti Marelli) are the only ones available today, and they are available in 40mm

Later, mid 1980s-on, 'Spansil' type float.

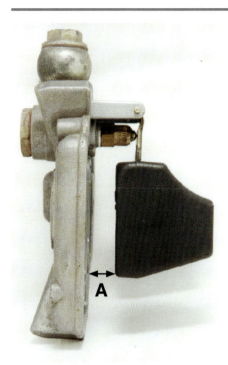

Fuel shut off height (A) with the top cover gasket in place is 12.0mm/0.475inch (the gasket is not in place in the photo).

none of them have fuel enrichment mechanisms fitted. The back of the carburettor is totally flat and as cast.

TOP COVER (WEBER)

The carburettor body is prepared next so that the top can be positioned and screwed on to it. The starter jet is

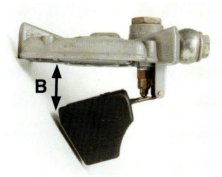

Float 'full droop' distance (B) is 26mm-27mm/1.025-1.065inch in the position shown with the top cover gasket in place (the gasket is not in place in the photo).

Late type carburettor body to top cover gasket.

screwed in (if it was removed), then the accelerator pump check balls and then the weights are placed in their respective holes. Then the screw-in plugs are fitted above these two items.

The accelerator pump control rod and piston are installed as a unit and the retaining plate snapped into position. The retaining plate can be squeezed together slightly using long-nosed pliers (but not too much

This Spanish built sidedraught SP Weber has 'idle by-pass' circuitry. The adjustment screw and lock nut is under the plastic cap arrowed.

as it will not spring back because it is made of brass) which have had their ends chamfered so that they fit the two indentations in the plate exactly. The plate can simply be tapped into position but this can damage the

Spanish built 48mm DCO/SP sidedraught SP Weber with flat back.

Weber spindle return spring anchor plate (arrowed) correctly fitted.

Weber accelerator pump intake valve (arrowed) fitted.

find out whether an increase in stroke works. If the increase in length proves to work, new rods of appropriate length can be ordered. Note that the ends of the rods are hardened to prevent wear. In most instances the standard rods can be made to work by changing the accelerator pump intake/discharge valve and accelerator pump jets, but, if the rod is too short or too long, it will have to be changed. On Webers, fit the shortest rod that does the job! An indication of the rod being too short is hesitation very late in the acceleration phase by way of the fact that the extra fuel progressively required during this period simply ceases to flow. Check the spring tension and condition and replace it if it seems weak or is corroded.

Fit the spindle return spring. The eye of the spring is fitted to the accelerator pump control lever first and the eye on the other end of the spring used to pull the spring up and through the hole where the anchoring plate is placed under the eye and the eye then let down into the register in the carburettor body. Care needs to be taken here when fitting the anchoring plate. The problem is one of holding the spring up high enough while fitting the anchoring plate. A small diameter shank screwdriver (2.5mm/0.100in diameter) with a hook bent on to its end (heated up, bent and then tempered back) is the solution and will prove to be fail-safe.

Fit the accelerator pump intake and discharge valve (if it has a hole drilled in the side of it) into the bottom of the fuel bowl. Do not over-tighten this valve as it can be very difficult to remove later. The slot size of the head is not very large and it is very easy to damage it.

aluminum of the carburettor body.

The stroke of the accelerator pump rod is measured for length of travel from the idle position to full throttle. Expect the stroke or travel to be 11mm (0.435in) on average. The stroke length will vary depending on the length of the rod. The longer the stroke (ultimately the length of the rod) the longer the duration of the

accelerator pump's shot of fuel. Unless the rod has been tampered with it will work well at 11mm stroke. It is unusual for the rod to be too short for most applications. Try the standard rods (meaning the ones that came with the carburettor) first as they can always be changed later if not satisfactory. If the rod is too short for the application, the ends can be built up with braze to

Weber baffle plate (arrowed) shown fitted. All is now ready for fitting of the top cover.

Make sure the baffle plate fits into its recess in the carburettor body.

The top is positioned on to the carburettor body and the five screws inserted into their respective holes and tightened progressively. One screw will have a tab under its head; this is the reference number for the particular carburettor. **Caution!** The floats are less prone to damage if the cover carrying them is installed as soon as possible after setting float heights.

Caution! Make sure that all of the screws used to secure the carburettor parts to the body have washers fitted between their heads and the body. This will prevent any marring of the aluminum of the castings.

BODY COMPONENTS (WEBER)

The main jet, emulsion tube, air corrector jet and emulsion tube holder combination are assembled and inserted into their respective holes and tightened using a screwdriver with a head that fits the slots exactly to avoid any possibility of damage to the slots.

The idle jet and idle jet holder are assembled and inserted into their respective holes and tightened then the jet inspection cover is placed over them and the wing nut tightened. Make sure there is a new gasket under the jet inspection cover.

The chokes first and then the auxiliary venturis are now inserted into the carburettor body and the retaining screw turned in until it firmly contacts the auxiliary venturi. **Caution!** On 45s and 48s the retaining screw that holds the auxiliary venturi must not be done up too tight as it will distort the auxiliary venturi. **Caution!** The retaining screws must be locked or they will come loose and fall out! Use the Weber clamping plate or lockwire

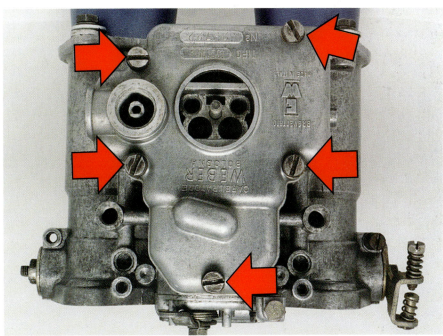

Weber top cover fitted. Arrows show securing screws.

the two screws together. Late model Webers use a screw and nut lock like Dellorto. The chokes and auxiliary venturis on the 40s push in and are retained and located by blade spring in a long slot machined into the throttle bore. They are not firmly fixed into position until the trumpets are bolted on or an air box is bolted on (carburettors supplied as original equipment to a car manufacturer).

Screw in the idle mixture adjusting screws with their springs and small washer fitted.

Weber emulsion tube holders (arrowed) fitted.

Weber idle jet holders (arrowed) fitted.

Screw in the progression hole plugs (new O-ring fitted).

The accelerator pump jets are inserted into their respective holes with the small aluminum washers in place. The screw-in plug is then fitted with a new O-ring seal fitted under the head of the plug.

The fuel union is fitted next. Fit new washers each side of the fuel union (sometimes called a 'banjo'). The fibre washer with the small hole in it goes against the carburettor body while the fibre washer with the larger hole goes against the union bolt. Do not tighten this bolt up at this point. Wait until the actual fitted position is known and always use a six-point box end wrench (ring spanner).

The filter inspection cover is fitted next. Fit a new filter element and a new fibre washer under the inspection plug. The fuel is under pressure and if the washer is not sealing well it will leak. Weber filters do not hold their shape so an inline filter should be fitted between the pump and the first carburettor fuel union.

Fit the small washer and then a cover to the end of the spindle and fit a

Right - Weber choke and an auxiliary venturi shown positioned in front of the carburetor before being installed.

Both 45mm and 48mm Webers use these set screws to secure their chokes and auxiliary venturis. Lock wire each pair of set screws through the holes drilled into them to prevent them coming out in service.

Weber idle mixture adjusting screws fitted.

new lock-washer before putting the nut on. The nut must be tensioned with the spindle firmly held. Hold the spindle adjacent to the throttle bore wall so there is less chance of twisting the spindle. The spindle is very strong but it is not impossible to damage it. Use long-nosed pliers (in conjunction with some shim to protect the butterfly) to hold the butterfly/spindle. Now check that at full throttle, when the throttle arm is on the stop, that the butterfly is

at 90 degrees to the throttle bore. If is isn't, replace the throttle arm with one that does allow the correct butterfly attitude, or build up the throttle arm's stop with braze and hand file it to suit.

The throttle arm is fitted next. The first thing to check is the fit of the slot of the throttle arm to the flat of the spindle (the round plain end covers are not critical here). If there is any looseness in this fit the throttle arm must either be replaced or repaired.

Weber progression hole cover plugs (arrowed) fitted.

Weber accelerator pump screw cover plugs (arrowed) fitted.

Below - Weber fuel union ('banjo'), washers and bolt shown in line as they should be fitted to the carburetor top cover.

It should be a tight fit on the spindle. Fit a new washer and tighten the nut. Use the long-nosed pliers once again to hold the spindle and bend the tabs over on both nuts when they are tight.

The bottom well cover is fitted next. With the carburettor turned upside-down, place a new gasket on the body and fit the well cover, then the four screws and tighten them.

At this point the Weber carburettor is assembled and ready to be fitted to the intake manifold.

Above - Weber fuel filter (arrowed) shown in position in its housing.

Later type of plastic and mesh Weber fuel filter. Early filters were just mesh.

Right - Weber plain cover (arrowed) and nut fitted to the end of the throttle spindle.

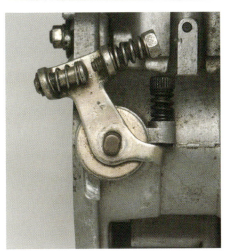

Weber linked-type throttle arm positioned on the throttle spindle. Note that it must be a tight fit on the spindle.

Weber bottom well cover components prior to fitting.

Weber bottom well cover fitted to the carburettor. Securing screws arrowed.

Chapter 3

Fuel management, air filters & ram tubes

FUEL FILTERS

Both carburettors are fitted with an integral filter. These filters should be left in place but, in the case of the Weber, they are often left out because the filter is not a particularly sturdy item and usually distorts the first time it is installed. The Dellorto system is better all round. With either carburettor set-up, install a paper element filter (throw away type) between the fuel pump and the first carburettor. These universal fitting fuel filters are available with 5/16in and 3/8in (or metric equivalent) intake and outlet diameters and they are cheap. This modification will ensure that only clean fuel is supplied to the carburettors. Note that steel cased in-line fuel filters are available and should be fitted when competition rules require it.

FUEL LINES (PIPES) & FITTINGS

Use only gasoline (petrol)-rated hose for fuel lines and, if you want the best, consider 'aircraft quality' metal-braided lines and companion fittings. The internal diameter of the standard fuel unions on these carburettors is 5.5mm (0.225in) which suits a 5/16in inside diameter fuel lines. The largest available fittings have an inside diameter of 7.5mm (0.30in) and suit 3/8in inside diameter fuel lines. Route fuel lines well away from the exhaust system and so that there are no sharp bends: secure well at frequent intervals.

Always use top quality hose clamps (clips) of the correct size for the application. The best clamps are the stainless steel worm-drive ones

A selection of in-line fuel filters.

A range of small clamps suitable for use on fuel lines (pipes). Use clamps of the correct size.

that close around the hose uniformly. Some of the larger types of clamp are too bulky to actually go down to the required size even though they are rated to do so.

AIR FILTERS

It is a proven fact that engines last a lot longer if the air admitted to the engine is as clean as possible. Good quality paper element filters are highly-efficient and usually feature rubber sealing between the filter and filter housing. Some air cleaner systems available (gauze and foam) do not offer good sealing between the filter and the filter housing body, so some unfiltered air is able to enter the engine. To reduce this possible problem, the filters can be sealed to their housings using silicone sealer (after testing that there is no reaction with the filter material). Always fit the very best air filtration system that you can afford to buy; it's money well spent.

Incidentally, wire mesh filters over ram stacks offer virtually no protection (except from stones) and can actually upset airflow considerably.

The most efficient filtering arrangements allow the fitting of ram tubes, a reasonable amount of internal airspace and a large good-quality paper air cleaner. The ideal may require the fabrication of an airbox using 3mm (1/8in) thick aluminum

Two 'socks' fitted to a carburettor. The 'socks' have been cut down to suit the application, but were left as long as possible in the circumstances.

This 'itg' foam air filter push fits over ram tubes.

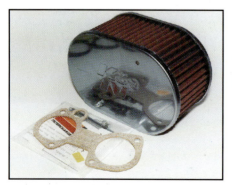

Compact K&N air filter assembly.

sheet. For a twin sidedraught application, for example, the airbox will measure approximately 400mm (16in) long by 140mm (5.5in) wide and a depth to accommodate the selected ram tubes plus 25mm (1in) clearance. If appropriate to your application, one of the best air filters to use is the original equipment item from the Jaguar XJ6 sedan (saloon). This filter is of the paper element type but of very sturdy construction with a rigid metal pressing all around. The filter can be fixed to the airbox by drilling a few holes around its perimeter and then bolting it into place. Remember to use silicon sealer between the filter and the airbox so that only filtered air can enter the engine.

There are many good proprietary designs of air cleaners available that can be considered 'bolt on' - provided they actually fit the installation. There are many different types of filtration material and all have their merits. With space requirements on most installations being limited, sometimes only one brand of filter will actually fit neatly. Use the filter or filters that fit and have the required airflow capacity (power ranges are given by filter manufacturers, so check the rating before purchasing). Prepare - and then rigorously maintain - the filter material as prescribed by the manufacturer.

When space is at a premium it may appear that there is hardly any space for ram tubes, let alone any form of air cleaner. However, it's better to run some form of good proprietary air cleaner than none at all. 'Socks' are available at reasonable cost, and they will fit nearly all applications. 'Socks' are pushed carefully over the short ram tubes and held in place using a plastic tie. The tie is pulled tight in the normal fashion, but not so tight as to really distort the filter element. Observe the manufacturer's recommendations for the preparation and cleaning of the filter element. Provided the filter is of sufficient size (not cut too short!) there will be no loss of power.

RAM TUBES (STACKS)

Ram tubes, trumpets or air horns (or whatever you know them as)

A selection of typical ram tubes.

Eurocarb made ram tubes for Dellorto and some Weber carburettors.

should always be fitted to a modified engine. There is a vast range of ram tubes available and recent products are usually efficient and scientifically designed. Modern ram tubes tend to be bellmouthed (the outer edge facing back to the carburettor). Ram

tubes are expensive, reflecting their manufacturing costs.

Ram tubes can be bought in various lengths, a factor which can be used to good effect on modified engines. It is worth experimenting with ram tube length to find the optimum for your particular application. As a general guide, short ram tubes are usually used for high rpm applications while longer rams are usually used to maximize mid-range response.

The ram tubes need to fit in the air filter housing with at least 25mm (1in) of intake clearance, so bear this in mind when making your choice unless you intend to fabricate your own housing or have a suitable proprietary brand in mind.

Over the years, some engines have been fitted with 'reversion plates', with good effect. A good example of this was the late 1960s Ford GT40 when equipped with the Weslake cylinder heads on a 302 cubic inch Ford V8 engine. The IDA Weber carburettors have what look like covers over the carburettor intakes but this is not what they're for, although it is true to say that they would prevent objects being dropped into the carburettors. These plates were fitted where they are (in relation to the ram tubes) after extensive engine dynomometer testing to position them where they do the most good. Most people don't go to this sort of trouble.

FUEL PRESSURE

Webers and Dellortos require high fuel volume not high fuel pressure. 1.5 to 2.5 pounds per square inch is the fuel pressure requirement. For testing purposes a fuel pressure gauge should be connected at the last fuel union (at the carburettor that receives the fuel last) and the fuel pressure monitored. Once it is established that the fuel pump can maintain the correct fuel pressure under all circumstances the fuel gauge can be removed.

To test the fuel pressure, run a pipe from the last carburettor in the fuel line through to the instrument panel and mount a gauge temporarily (or permanently) within easy view

Late model plastic-bodied SU fuel pump.

so that the fuel pressure can be monitored.

If the fuel pressure is going to fall and cause problems (starvation) it will usually do so when the engine is at full speed in top gear after a short distance has been covered. What happens is that the fuel pump does not keep up with the fuel demand of the engine under these load conditions and the pressure simply drops. The end result is a hesitation as the fuel level drops. If you have a gauge installed, as soon as the pressure goes you can stop. Do not proceed with high rpm testing until the correct fuel pressure can be maintained, otherwise the engine could be seriously damaged. Once the fuel pressure situation has been thoroughly checked out, the fuel pressure gauge and lines can be removed.

High-performance Facet fuel pump.

Achieving a fuel pressure of at least 1.5psi but not more than 2.5psi when the engine's fuel consumption is at its maximum is the aim. Many standard mechanical fuel pumps will keep up with the fuel requirements of Webers and Dellortos. If the fuel delivery system is in first class order but still does not deliver enough fuel, you'll need to consider fitting a mechanical pump of higher capacity than the current item. It is also possible to bypass the standard mechanical pump and use instead a larger capacity electric pump to deliver the fuel.

If a single electric fuel pump is fitted as standard and a new one cannot supply sufficient fuel, consider fitting a second electric pump of the same type in parallel with the first. The second pump must have its own pick-up line (pipe) from the tank. The outlets of the two pumps can be merged into a single fuel line with a minimum inside diameter of 3/8in (or metric equivalent). The larger line can be reduced to 5/16in just before the fuel intake of the first (or only) carburettor so that the standard carburettor unions can be used.

Large capacity single electric pumps are available from several manufacturers (Holley and Facet for example) and these can pump at up to 10psi and more, so they'll have to be used with a fuel pressure regulator. Too much pressure can overwhelm the needle valves and cause over-rich running.

Chapter 4

Choosing the components for your carburettor/s

COMPONENTS - INITIAL SELECTION

The following tuning information is designed to assist you find a suitable start-up and run specification for your carburettor/s and your application. The set-up may not be perfect, but it will be quite sufficient to start with and can be improved upon by fine tuning.

Weber and Dellorto parts are high priced and usually quite a few of them are needed to fine tune an engine. Invariably parts are borrowed from other people to see if they work and then new or second-hand parts bought for final fitting to the engine. It can get very expensive having to buy jets just to try them out. Engine tuners usually (but not all) have the majority of parts in stock by way of a 'tuning kit'.

When tuning your own engine the cost of buying parts has to be balanced against the actual cost of getting a specialist firm to carry out this work. It is possible to spend a considerable amount of money on jets, chokes, etc.,

(new or second-hand) that you will end up not using. Conversely, if the firm you are getting your engine tuned at does not have a comprehensive set of alternative parts the chances are they will not tune the engine correctly. They might get it going better than you think you can, but that doesn't mean that it's actually right.

CHOKE SIZE VERSUS CARBURETTOR SIZE

The information given here will allow you to sort out any sidedraught Weber or Dellorto on any suitable engine application. The first consideration before carburettors are purchased is to know what choke sizes are likely to be used. The reason for this is that carburettor bodies of a given size have a range of chokes that will fit them. The correct range of choke sizes for both Weber and Dellorto carburettors is as follows.

40mm diameter carburettors:

choke range 28mm-34mm
45mm diameter carburettors:
choke range 36mm to 41mm
48mm diameter carburettors:
choke range 41mm to 43mm

There is some overlap of possible choke sizes on these carburettors. A 34mm choke is definitely the maximum size to use in a 40 DCOE or 40 DHLA body. 36mm chokes are available from Weber and Dellorto for fitting into their 40s but the flow capacity is definitely not as good as a 36mm choke in a 45 body. The 45s can have 34mm chokes fitted which, while within the sizing tolerance of a 40 body, still work well in a 45. The 48s could have 34mm chokes fitted (in the range of 40 bodies) or 36mm, 38mm and 40mm chokes fitted (in the range of 45 bodies) and also work well.

Choke size is the major consideration. Very small choke sizes in large carburettor bodies is not a good idea so consider 34mm chokes to

Dellorto chokes, 40mm version on the left, 45mm centre and 48mm right.

Weber chokes to suit a 45mm carburettor.

be the smallest size for 45s and 36mm chokes the smallest size for 48s. The range of choke sizes which are readily available for the most widely used carburettor bodies goes up in 2mm increments as follows.

40s: 28, 30, 32 or 34mm chokes.
45s: 34, 36, 38 or 40mm chokes.
48s: 36, 38, 40 or 42mm chokes.

The choke size range for each model of carburettor is quite large as they go up in small increments, this means the choke size can be closely matched to the needs of the engine. Most of the sizes are listed here including those of the 38mm version of the DCOE -

38s: 26, 27, 28, 29, 30, 31, 32mm chokes.
40s: 28, 29, 30, 31, 32, 33, 34mm chokes.
45s: 34, 35, 36, 37, 38, 39, 40, 41mm chokes.
48S: 36, 37, 38, 39, 40, 41, 42, 43mm chokes.

CHOKE SIZE - SELECTING

In the first instance it is essential to know what each carburettor will take with regard to choke sizes and that is all listed in the previous section. The next stage is to decide on the choke size required for your particular engine and this is done on the basis of the individual cylinder capacity and the rpm range that you are likely to use.

As an overall recommendation

you are advised to FIT THE SMALLEST CHOKE THAT WILL GIVE FULL POWER.

When a slight choke size reduction is necessary to achieve better low end performance (eg: out of a corner pulling power) then FIT A CHOKE SIZE THAT PROVES BEST FOR THE OVERALL APPLICATION. As an example, an engine may well produce most maximum rpm power with 38mm chokes but, because in reality the engine is usually used over a wide rev range, 36mm chokes will prove better all round, offering superior mid-range with only a slight loss at the top end. Testing and (golden rule number one) changing one thing at a time is the only way to find out which is the best overall solution for you.

Most engines that Webers or Dellortos are fitted to are modified in some way and the degree of modification will have some bearing on the choke size that will work best. The situation is not as bad as it sounds, and choke choices can be narrowed down. Be realistic on how good your engine actually is and how fast you actually are going to turn your engine on a regular basis. Road cars should be fitted with the smallest chokes possible which are conducive to good all round engine performance. Jetting should be set with good emissions in mind without being excessively lean. Webers and Dellortos can give very good economy coupled with good performance but it is also fair to say that generally they will use more fuel than the original carburettor. For a start, in most applications, there's an accelerator pump for every cylinder.

A well modified four-cylinder 2000cc or 2100cc engine fitted with twin sidedraughts will usually run best using 38mm chokes with 40mm chokes being too large. On the other side of the scale, if mid-range performance is preferred to absolute top end power 36mm diameter chokes will prove to be a better choice. The readily available range of choke sizes increases in 2mm increments and is nearly always satisfactory for achieving the 'right' size.

It is reasonably easy to establish the range of choke sizes and then know what carburettor you will require. If the engine needs 36mm or 38mm chokes then 45 carburettors are the size to buy. If the engine needs 34mm chokes then 40 carburettors are usually the size to buy but this choke size is at the limit of the 40 carburettor body. The latter is a good example of the choice of carburettor body versus the true requirements with regard to choke size. In most instances, unless your engine really is a heavily modified unit, the 40s with 34mm chokes will prove to be the best choice, especially for a street car.

The following recommended choke sizes are for engines that are modified and have an effective rev range from approximately 3000rpm to 8000rpm for the smaller engines (1000cc to 1750cc) and 2500rpm to 7500rpm for intermediate engines (1750cc to 2400cc) and around 2000rpm to 6500rpm for larger units.

Many engines are bored out beyond standard capacities; however, when consulting the following choke size tables, use the actual capacity of your engine.

For your convenience the following choke sizes are listed in relation to ccs per cylinder so that the sizes can be cross referenced to any engine. If your engine is an in-between capacity go to the next choke size down. Always be prepared to go to a smaller choke size if tuning proves difficult. The common choke sizes listed below are basic starting points.

Choke sizes (on basis of one choke per cylinder)-

28mm chokes: 250cc per cylinder.
30mm chokes: 300cc per cylinder.
32mm chokes: 350cc per cylinder.
34mm chokes: 400cc per cylinder.
36mm chokes: 462cc per cylinder.
38mm chokes: 525cc per cylinder.
40mm chokes: 600cc per cylinder.
43mm chokes: 800cc per cylinder.

The even sized choke numbers (32, 34, 36 and so on) are far more common and easily obtained than the odd sized choke numbers (31, 33, 35 and so on) and, as a consequence, engines are generally fitted with the most suitable even numbered sized choke. A 1300cc engine can have 30mm choke fitted initially but the engine may prove to need 32mm (or even 31mm) chokes after testing. The chart above is close enough for initial setting and the 2mm increments in choke sizing not usually too large as steps up or down during fine tuning. There is a degree of latitude with choke sizes but one choke size will always work better than all of the others for a particular application.

In the first instance use the even numbered chokes simply because of their easy availability but be aware of the fact that in-between sizes do exist and an odd sized choke may ultimately prove to be the correct one for your engine.

Intermediate choke sizes (on basis of one choke per cylinder) -

29mm chokes: 278cc per cylinder.
31mm chokes: 325cc per cylinder.
33mm chokes: 375cc per cylinder.
35mm chokes: 425cc per cylinder.
37mm chokes: 475cc per cylinder.
39mm chokes: 565cc per cylinder

The list which follows is in addition to the above and for large capacity engines.

Choke sizes for large engines (on basis of one choke per cylinder) -

40mm chokes in 45s or 48s: 600cc per cylinder.
41mm chokes in 45s or 48s: 700cc per cylinder.
42mm chokes in 48s: 800cc per cylinder.
43mm chokes in 48s: 900cc per cylinder.

There is another scenario that can be applied to all engines that are modified for occasional competition yet must be used most of the time as daily transport. That is to have two sets of chokes and jets: one set for each situation. If, for instance, a 2000cc four-cylinder engine with one choke per cylinder needs 38mm chokes with suitable jetting for competition use to obtain a maximum usable rpm of 7800, there's no reason why these chokes cannot be changed later to 34mm with suitable jetting. The low end performance of the engine will actually be improved and all you will notice at the top end is a very sudden and definite flattening off at a certain rpm, beyond which the engine simply will not go. With smaller chokes fitted the jetting can be reduced without detriment. What you will have done is governed the maximum engine speed by restricting the air supply. With 34mm chokes fitted, a good 2000cc engine will usually still go well up to 6000rpm or so.

Some engines, notably the BMC/Rover A-series and B-series four-cylinder engines, are Siamese intake port units. A single sidedraught is usually fitted to these units, each choke feeding two cylinders. The choke selection for these engines can be based on the following -

Chokes for A-series (one sidedraught) -

850cc: 30mm chokes in 40mm carburettors.
998cc: 32mm chokes in 40mm carburettors.
1098cc: 34mm chokes in 40mm carburettors.
1275cc: 36mm chokes in 45mm carburettors.
1380cc: 38mm chokes in 45mm carburettors

Chokes for B-series (one sidedraught) -

1800cc (mild state of tune): 36 or 37mm chokes in 45 carburettor.
1800cc (well modified state of tune): 38 or 39mm chokes in 45 carburettor.
1900cc (bored out, well modified state of tune): 40 or 41mm chokes in 45 carburettor.

Note that all of the above choke size recommendations are average sizes to be used as guides. The fitting of smaller chokes than recommended will not automatically result in a gross power loss, far from it, in fact, in many instances a particular engine will go better with smaller chokes. For example, an 850cc A-Series engine will go extremely well with 28mm chokes. The suggestion is really to try whatever chokes you have at hand which, while being smaller than recommended, are not outrageously smaller. Smaller chokes often result in a better accelerating engine with a slight reduction in top end power. If most of your work is not all out top end then you could well be better off with smaller chokes and jetting to match.

IDLE JET - SELECTING

Idle jet selection, or progression jet selection as it is often known, is based on the smallest jet size being used (compatible with good idling and, most importantly, good progression), balanced against the most suitable air bleed hole being used to give the correct mixture. A richer than necessary mixture will cover the progression phase but it will be wasting fuel and frequently causing the engine to accelerate slower than it would if the jets were sized correctly. If all jetting was done on the basis of the richer the better there would be little economy and the overall performance would be very poor. Over rich fuel mixtures do not produce optimum power.

On all Dellortos and Webers the idle jet system feeds an air/fuel mixture to the idle mixture adjusting screw. The idle mixture adjusting screw controls the amount of air/fuel mixture admitted to the engine for idling purposes but the idle jet combination controls the overall ratio of that air/fuel mixture. The idle mixture screw is passing an air/fuel mixture which, when entering the throttle bore, is further mixed (leaned off by dilution) with the amount of air passed by the butterfly in its slightly open state.

The idle jet chart gives reasonable

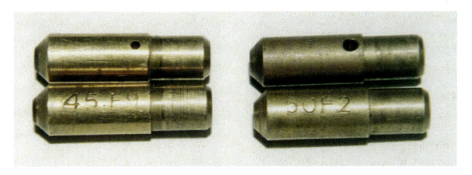

Numbers indicating the fuel hole size, followed by a letter/number combination indicating the air bleed size on Weber idle jets.

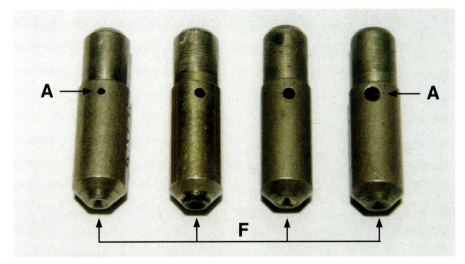

Numbered fuel component of the idle jet is the size of the hole drilled in the end of the idle jet (F). Letter/number air bleed component of the idle jet is hole size in the side of the jet (A).

first number starting point sizes for idle jet selection. It is unlikely that the idle jet would need to be larger than the sizes listed, and consider five sizes down from the given listing to be the smallest ever likely to be needed. The final sizing of the idle jet is only known after the progression phase of the engine is tested because, although the engine can be set to idle with a certain sized jet installed, that does not mean that the same jet will give good progression, especially under load. The idle jet and air bleed size is selected finally on the combination's ability to give good progression (from idle to main jet operation).

Sorting out an idle jet for idling purposes only may seem like double work as the jet combination may well have to be changed at the next stage of the tune-up. This is a correct assessment of the situation but it is helpful if the engine is able to idle well, before the progression phase is checked. It is easier if only one thing is being sorted out at a time.

Idle jet codes (Weber)
Weber uses the same idle system

Here the bleed hole is right at the top of the carburettor body and is not recessed 8mm, or so, as on normal DHLAs.

Normal Weber carburettor idle jet holder on the left and emission type suitable only for use in emission carburettors on the right.

throughout the DCOE range. The idle jets are coded 45F6, 55F2, 60F11 and so on. The fuel component of the range is stated by the first two numbers 35, 40, 45, 50, 55 etc. and denotes the hole size in hundredths of a millimeter and the sizes go up in 5 hundredths of a millimeter increments. All Weber idle jets have a suffix such as F6, F9, F11, F2, and so on, which denotes the air bleed sizing or hole sizing with F6 having the smallest hole and richest mixture strength. The air bleed hole is on the side of the jet. There are two parts to the numbering and lettering on Weber idle jets and the fuel hole and the air bleed are on the same jet.

It is possible to have 45F9, 45F6, 45F2, 45F8, and so on, combinations. This means that the fuel component of the idle jet (the hole in the taper seated end of the jet) is the same in all jets but the size of the air bleed holes varies. A small air bleed hole means a richer mixture and a larger air bleed hole a leaner mixture. A considerable amount of mixture alteration is possible with this system.

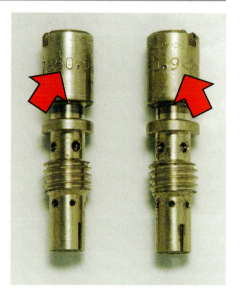

Dellorto idle jet holder. Note the positioning of the air bleed holes which control the idle mixture and the progression phase mixture strength. Size numbers arrowed.

Emission controlled 40mm sidedraught Weber carburettors
These carburettors have their very own idle jet holder which has a neoprene 'O' ring fitted at the top instead of a second thread. They use the standard Weber idle jet which has the air component (certain hole size) and the fuel component (certain hole size) as per all other sidedraught Webers. What is different about the way these carburettors work is the fact that the fuel that is to go through each idle jet has to pass through the adjacent main jet first. The thread cut into the body of the carburettor starts 7.5mm down the hole, rather than at the top of the idle jet hole. A conventional idle jet holder will screw in but is not 'O' ring sealed and won't work.

Idle jet codes (Dellorto)
Be aware that emission type Dellorto DHLA 40F, G, H, L and N models have only one idle jet holder and it has a top row of holes only (that's

above the threaded section of the holder and before the 'O' ring groove). These holes pass the fuel into the small mixing chamber which leads on into the carburettor throttle bore. The air bleed size is fixed by the size of the hole (2.00mm diameter) drilled into the actual body of the carburettor. This leaves only the fuel jet to be changed and fitted with much larger idle jets than normal. These specially built emission type carburettors can, in certain circumstances, prove difficult to tune when fitted onto other types of engine, especially very small capacity engines. Most engines fitted with these emission type carburettors will, however, respond well to the fitting of a large idle jet, such as a 60 or a 65, even on a small capacity engine. This is the general cure for off idle and progression hesitation problems. The problem is the fixed sizing of the air bleed. Avoid these carburettors if complete adjustability is required for your application. Larger four-cylinder engines (1750 to 2000cc) will tune well with these carburettors, it is the smaller cc per cylinder engines that will be too lean just off idle as the engine tries to go through the progression phase.

If fitting large idle jets doesn't work satisfactorily, you could try having small headed sleeves made with a nominal outside diameter of 2.00mm and an inside diameter of 1.2mm. These precision sleeves, which you will have to have custom made, are then light press fitted (tap in fit) into each of the bleed holes (two per carburettor). This will make the air bleed component the same as a 7850.2 idle jet holder.

The new hole (1.2mm) can always be opened out in 0.1mm increments for correct air bleed sizing. With the size of the pre-drilled air bleed being quite large, these carburettors have more chance of being successfully tuned without the need for custom

made supplementary jets, so long as the engine has at least 500cc per cylinder capacity. However, this still leaves the possible problem of the number and size of the progression holes to contend with (i.e., the carburettor is not really suitable for the application). It's only the DHLA 40 'universal performance' 40mm carburettors that are totally tuneable, virtually irrespective of the application.

Note that the universal performance 40mm sidedraught Dellorto only has DHLA 40 cast on it, with no stamped on letter.

The DHLA 40, DHLA 40C and the DHLA 40E marked carburettors are the ones to have for high performance or racing applications. They are fully adjustable and can have small air bleed sizes in the idle jet holder. The DHLA 40E type Dellorto is non-emission, and very similar to the C model, but has idle mixture adjustment screws in towers which can be made tamper-proof once adjusted. All three are 'universal performance' types.

On the 'universal performance' DHLA 40 carburettors, and the 45s and 48s, the idle jet holder has the air bleed hole (s) in it and there is a separate idle jet (for fuel metering). Effectively this is very similar to the Weber arrangement. The holder has two sets of holes in it: one set to allow the air/fuel mixture to enter the carburettor body (the highest set) and

Universal performance DHLA 40 has the letters and numbers cast in as shown in this photo.

another to allow the air to mix with the fuel (lower set). The lower hole/s range in size to allow more or less air to enter the central portion of the holder and then mix with the fuel.

There are ten code-numbered idle jet holders (7850.10 down to 7850.1) but they are not numerically in sequence going lean to rich. The 7850.1 idle jet holder, for example, has one 1.4mm diameter hole in it while the 7850.8 idle jet holder has four 0.5mm holes in it (giving a leaner mixture with the same sized fuel jet).

The Dellorto fuel jets go up in 1s (40, 41, 42, 43, 44, 45 and so on) which gives a huge range to choose from. Expect to make size changes in increments of 2 or 3 (40, 42, 45, 48 and so on).

Weber main jets (left) and Dellorto main jets (right) showing the position of the sizing numbers.

Idle jet and air bleed component - selection

The approximate idle jet sizes (Dellortos and Webers) for different applications which follow, unless otherwise stated, are based on one choke per cylinder applications. While Dellorto's jet range is far more comprehensive, the numbers of both company's jets pertain to the size of the holes in hundredths of a millimeter so with Weber the difference in hole diameter between a 45 and a 50 is 0.05mm (0.0025in). Dellorto idle jets go up in 0.01mm (0.0005in) metric sizes.

40 idle jet: 250cc to 350cc per cylinder.
45 idle jet: 350cc to 420cc per cylinder.
50 idle jet: 420cc to 490cc per cylinder.
55 idle jet: 490cc to 560cc per cylinder.
60 idle jet: 560cc to 630cc per cylinder.
65 idle jet: 630cc to 700cc per cylinder.

With the idle fuel jet selection made, that still leaves the air bleed component of the combination to be chosen. The choice, for performance engines, can be narrowed down going lean to rich to F2, F11, F8, F9 and F6 for Webers and usable steps 7850.1, 7850.6, 7850.7, 7850.2 and 7850.8 for Dellortos.

Regardless of engine capacity, initially start with the leanest of the following (F2 the leanest in the Weber range and 7850.8 the leanest in the Dellorto range).

The fuel and air bleed sizes given are approximations only and optimum sizes can only be found by testing the combinations. Refer to chapter 6 for details on how to narrow

down the jet size so that the leanest mixture compatible with correct engine performance is obtained.

The air bleed is the cross-sectional area (by way of holes) available for an amount of air to be admitted to pre-mix with the fuel which then becomes the air fuel mixture that is fed to the idle mixture adjusting screws and also, most importantly, the progression holes which are positioned downstream of the butterfly as it opens.

Weber idle jets with the same letter and number suffix on them all have the same sized holes in the side of the jet. That is what the code means. The fuel drilling will vary and be 45, 50, 55 etc. So each idle jet size in the F9 range for example produces a different idle mixture through having a different fuel intake hole size designated by the 45, 50, 55 prefix number.

The complete range of Dellorto idle jet holders are numbered 7850.1, 7850.2, 7850.3, 7850.4, 7850.5, 7850.6, 7850.7, 7850.8, 7850.9 and 7850.10 which suit all DHLA non-emission 40mm, 45mm and 48mm sidedraught Dellorto carburettors. What categorizes each idle jet holder on the basis of lean to rich is the size and the number of holes in the lower part of each idle jet holder. The general order of idle jet holders on the basis of going lean to rich for any given idle jet is: 7850.10, 7850.5, 7850.9, 7850.4, 7850.1, 7850.3, 7850.6, 7850.7, 7850.2, and 7850.8. The 7850.1 idle jet holder, for example, has one 1.4mm diameter hole in it, while the 7850.8 idle jet holder has four 0.55mm diameter holes in it giving it a leaner mixture with the same sized fuel jet.

The Dellorto air bleed is used to fine tune the mixture. This, in principle, is identical to the Weber but in the case of the Dellorto the idle jet holder itself is changed as this component has the

air bleed holes in it. With the idle jet selected from the chart given, the air bleed factor is increased or decreased by changing the idle jet holder.

Immediately below is the complete range of DCOE idle jet air bleed suffixes listed in order lean to rich -

F3, F1, F7, F5, F4, F2, F13, F11, F8, F9, F12 and F6.

Note that when a single sidedraught carburettor is fitted to a four cylinder engine (BMC/Rover Siamese intake port A-series engines, for instance), expect to use 40 idle jets on engines up to 1100cc capacity, and 45 idle jets on larger versions. For the air bleed component, the F9 will almost always prove correct for Weber, with the 7850.1 or 7850.2 proving correct for Dellorto carburettors.

Idle mixture and progression holes

The idle jet supplies air/fuel mixture to the progression holes. The passageway that takes the air/fuel mixture to the idle mixture screw passes over a series of holes that, when the throttle is at the idle speed opening, are positioned upstream of the butterfly. As the butterfly opens further these holes are swept over by the edge of the butterfly and are then subjected to engine vacuum as they each become downstream of the butterfly. An accelerating engine needs a richer fuel mixture and on these carburettors it gets one instantly by way of the fact that the air/fuel mixture is already flowing to the idle mixture adjusting screw, and when the throttle is opened the air/fuel mixture simply drops in through the holes as they are exposed to vacuum.

The point is that, if the engine is to accelerate smoothly, the air/fuel mixture that goes in through these holes has to be exactly right. This is why the idle jets have to suit the progression phase perfectly. The air fuel mixture for idling has a degree of adjustment via the idle mixture adjusting screw but the progression mixture can only be altered by changing the jets for an overall mixture change.

The progression holes get their air/fuel mixture from the idle jet, as do the idle mixture adjusting screws. The idle jet supplies both the idle mixture adjusting screw and the progression holes from the same passageway. The air/fuel mixture is already flowing to the idle mixture adjusting screw and, as the progression holes are exposed (by opening the throttle), the mixture simply falls in instantly and at a suitable strength. If the mixture going to the idle jet is a very lean one it will not assist the engine in the acceleration phase as it is designed to do and will cause hesitation.

If the throttle is opened too much at idle the first of the progression holes will be in operation: this should never be necessary and is not desirable. The engine has to idle without any progression hole assistance (extra air/fuel mixture). From idle to main jet operation the progression holes fill what would otherwise be a flat spot by instantaneously adding air/fuel mixture to the main air stream. Just as the name suggests these holes are progression holes used to smooth the way between idle and main jet operation.

IDLE SCREW ADJUSTMENT PROCEDURE

The idle screw adjustment procedure is the same for both carburettors. Turn the screw in to lean the mixture off and turn it out to richen the mixture. The idle screws themselves vary quite a bit. The Weber uses a fairly coarse thread

which is technically quite correct for something tapped into a soft material such as an aluminum carburettor body. The later Dellorto carburettor uses a very fine pitched thread which is not really good practice when used in cast aluminum. The spring and the idle screw is enclosed in a tower. The earlier Dellortos had coarse threads and are very similar to the Weber. The Weber will characteristically require 7/8 to 1 1/2 turns out of the idle adjustment screw to effect smooth idle, while the later fine thread idle mixture adjustment screw Dellorto will usually require 5 to 6 1/2 turns out from the fully seated position. Of course, the objective is to find a position - by progressively turning the screw outwards - which allows the engine to accelerate from idle speed without hesitation. A slightly rich idle mixture setting can be used to advantage to supply excess fuel for initial acceleration purposes.

MAIN JET - SELECTION

Both carburettors have a numbering system for the main jets but, whereas Weber uses 135, 140, 145, 150 and so on, Dellorto uses 135, 136, 137, 138, 139, 140 and so on. A considerable range of sizes is available from both companies but the Weber size increments are usually satisfactory, Dellorto's bigger range via smaller incremental increases is a very useful feature. For example, with a Weber the main jets may be 140 and not rich enough for the application so 145 will have to be fitted; however, this size increase may be slightly more than is necessary. With a Dellorto, going from 140 to 142 may well prove to be exactly right. It's a small point, but it works in favor of Dellorto.

The following chart gives main jet sizes for modified engines with choke sizes chosen as indicated earlier. The

recommendations represent the rich side or 'jetted up' side for the given applications so jetting down may be necessary for optimum results - but only after testing. The main jet must be sized correctly because over-rich mixtures wash the oil film off the cylinder bores and cause high cylinder wear; conversely, lean mixtures can damage the engine through overheating. For mildly tuned engines, expect to drop up to 5 in jet size from the following recommendations -

Main jet recommendations with ideal choke size (one choke per cylinder applications) -

120: 250cc per cylinder.
125: 300cc per cylinder.
130: 350cc per cylinder.
135: 400cc per cylinder.
140: 450cc per cylinder.
145: 500cc per cylinder.
150: 550cc per cylinder.
155: 600cc per cylinder.
160: 650cc per cylinder.
165: 700cc per cylinder.

Siamese intake port engines from the BMC/Leyland range should work with the following main jet sizes.

Main jet recommendations for Siamese port engines -

30mm choke: 135 main jet.
32mm choke: 145 main jet.
34mm choke: 155 main jet.
36mm choke: 160 main jet.
38mm choke: 165-170 main jet.
40mm choke: 175-180 main jet.

EMULSION TUBE - SELECTION

The emulsion tubes affect the acceleration phase of the engine as the main jet circuit comes into operation, during further acceleration

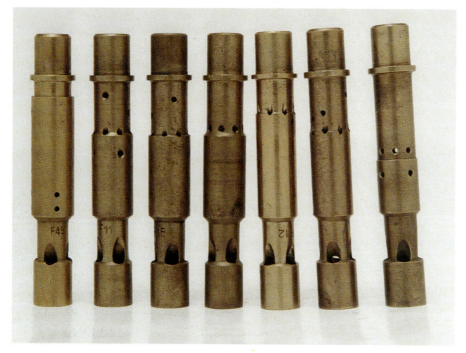

An assortment of Weber emulsion tubes showing the positions of the small holes, the diameters of the central sections of the emulsion tubes, and the different lengths of the central sections.

Weber emulsion tube (left) and Dellorto emulsion tube (right). Position of sizing numbers arrowed.

with the main jet circuit in operation, and up until nearly maximum rpm. If the emulsion tubes are not correct for the application the engine will not accelerate cleanly or quickly (incorrect air/fuel mixture).

Emulsion tubes are located precisely in the body of the carburettor by the accuracy of the componentry (jets, emulsion tube and air corrector) and the way the body of the carburettor is machined (these carburettors and their parts are made to extremely fine tolerances).

It's also essential that the floats are set precisely so that that the level of the fuel in the emulsion tube well is as prescribed by the carburettor manufacturer. Fuel level accuracy is critical in relation to the small holes in the upper part of the emulsion tube. It is, for example, possible to have the fuel level so high that some of the small holes of some emulsion tubes are

covered when they shouldn't be. This will result in an extremely rich mixture until the level is reduced, and is one of the main reasons why fuel levels are so important.

There is a great deal of variety when it comes to emulsion tube types: the diameter of the emulsion tubes varies, the step height varies, the position of the small holes varies, the number of small holes varies, and the diameter of the holes varies. Some emulsion tubes have small holes positioned quite high up, while others have one or more small holes lower down, or positioned midway, to allow the mixture to be weakened as the reservoir level drops when the main circuit comes into operation. It's all about obtaining the correct air/fuel delivery curve for optimum acceleration smoothness as the main circuit comes into operation.

Fuel always enters the emulsion

tube well through the main jet and the four holes on the lower part of the emulsion tube. Weber emulsion tubes have four steeply-angled holes, while Dellorto tubes have holes drilled at 90 degrees to the main axis. Air enters the emulsion tube through the air corrector, and then passes through the small holes in the middle/upper portion of the emulsion tube into the well, where it mixes with the fuel. The emulsified air and fuel mixture then passes into the auxiliary venturi and on into the engine.

The larger the diameter of the emulsion tube (the central portion), the smaller the reservoir of fuel in the emulsion tube well. Conversely, the smaller the diameter of the emulsion tube the larger the reservoir of fuel in the emulsion tube well. The larger the reservoir of fuel in the emulsion tube well, the more fuel there is available for the acceleration phase, and when the main circuit starts to operate. If there is too much fuel in the reservoir for the particular engine, the mixture will be too rich, and if there is insufficient fuel in the reservoir there won't be enough fuel instantly available for the smooth transition to the main circuit.

With the engine at or near maximum rpm under wide open throttle conditions, the emulsion tube well is under depression/vacuum (no reservoir of fuel remaining in the emulsion well). The fuel coming in through the main jet goes through the four holes in the lower part of the emulsion tube, and mixes (emulsifies) with the air which comes from the air corrector through the small holes drilled in the upper part of the emulsion tube. The size of hole in the main jet controls the amount of fuel that can go into the engine in much the same way as the air corrector controls the amount of air that can enter the emulsion tube. Once in the centre of the auxiliary venturi, the air/fuel mixture passes into the main air stream and is mixed with the main body of air drawn into the engine.

The reason why the air corrector holes size has an effect at high/maximum rpm is because it controls the maximum amount of air that can be drawn into the emulsion tube well to be pre-mixed with the fuel. At high rpm, the tendency is for the air to flow through this circuit at a faster rate than at lower rpm, and, as a result, the mixture weakens (proportionally less fuel is drawn into the auxiliary venturi). This would result in a high speed hesitation, and would require a reduction in the size of the emulsion tube (220 to 200 or 190 to 180, for example) to increase the fuel content.

When the engine is at maximum rpm and wide open throttle, there is no

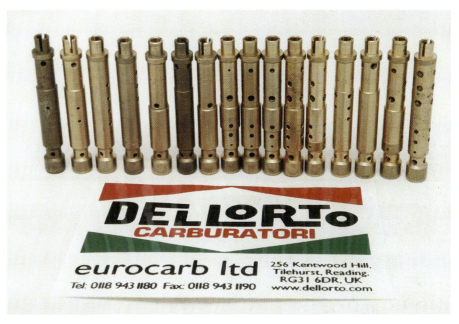

The complete range of Dellorto emulsion tubes from 1 to 16, left to right.

The 7772.4 tube is on the left and the 7772.3 tube is on the right.

reservoir of fuel in the emulsion tube well, there is only fuel and air passing through the void, into the auxiliary venturi and into the engine. Once the throttle has been closed, the emulsion tube well will fill with fuel again, to the level of the fuel bowl, in readiness for the next repeat cycle.

The emulsion tubes of both carburettors are graded according to their central section diameters, the depth of the central hole from the air corrector end of the emulsion tube, and the number, diameter and position of the small holes in the central section. The most commonly used Weber emulsion tubes, going from lean to rich are: F11, F15, F16, F2, F8 and F7. These emulsion tubes will cover most applications.

Consider F15 to be the starting point emulsion tube on engines which have 250cc to 325cc per cylinder capacities and one choke per cylinder; F16s as the starting point emulsion tubes on engines that have 300cc to 425cc per cylinder capacities and one choke per cylinder; and F2 emulsion tubes as the starting point emulsion tubes as starting points on 425cc to 550cc per cylinder capacity and one choke per cylinder engines. This is the usual range to consider but, of course, there are other Weber emulsion tubes (F1, F3, F4, F5, F9, F10, F12, F14, F17, F20, F21, F22, F24, F25, F26, F27, F28, F30, F32, F33, F34, F36, F38, F39, F41, F42, F43, F44, F46, F47, F49, F50). These aren't commonly used on single, twin or triple sidedraught Weber applications, which are the most common, and you don't really need to concern yourself with them.

If you have F15 emulsion tubes fitted, and the engine has hesitation in the 1600-2000rpm to 2600-3000rpm acceleration phase as the main jet system comes into operation, or is in operation, try changing the emulsion tubes to F16s. The use of F8 and F7 emulsion tubes is not normally necessary on sidedraught applications but could be looked at if F2 emulsion tubes don't prove to be enough. The usual application that would see F8 and F7 emulsion tubes used is a large capacity engine (with 600cc per cylinder capacities and above) which requires low rpm richness when the main jet system starts to operate. Realistically, F2s are almost always enough.

Dellorto emulsion tubes follow a numerical sequence from 7772.1 through to 7772.16, though this should not imply a progression from lean to rich, for example. Dellorto emulsion tubes can be divided into three distinct groups, with the first comprising the 7772.3 and 7772.4 emulsion tubes. These were made for DHLB32/35 carburettors and, although they will fit, are not in our line up for all round DHLA use.

The next group comprises 7772.8

through to 7772.16 which were for use on emission control carburettors. The third group which overlaps slightly into the previously mentioned emission control carburettor group is that used for all general performance applications of 40mm, 45mm and 48mm Dellorto carburettors. This last group is listed here on the basis of going lean to rich for your convenience: 7772.10, 7772.8, 7772.2, 7772.1, 7772.14, 7772.5, 7772.7 and 7772.6.

7772.5 is no longer available from Dellorto, although there are literally thousands of these emulsion tubes available on the secondhand market. 7772.14 is an emission control carburettor emulsion tube, but this is where it fits in the overall scheme.

Consider 7772.10 or 7772.8 emulsion tubes as starting points on standard engines from 250cc up to 500cc per cylinder capacity and one choke per cylinder, and 7772.2, 7772.1 and 7772.14 emulsion tubes as the next steps up. The latter should also be considered as starting point emulsion tubes for modified small capacity per cylinder engines (250cc to 400cc per cylinder) with one choke per cylinder.

The final three emulsion tubes (7772.5, 7772.7 and 7772.6) are the most frequently used for modified and racing engines with cylinder capacities from 300cc through to 700cc. Consider the 7772.5 emulsion tube as the starting point for 300-500cc per cylinder modified engines, and 7772.7 emulsion tubes for 400cc to 500cc per cylinder engines with one choke per cylinder. Use 7772.6 emulsion tubes on racing engines with 500cc per cylinder capacities and up with one choke per cylinder. Use the leanest of these three emulsion tubes that gives your particular engine the best all round performance.

Siamesed intake port engines, as

Weber air corrector (bottom left) and emulsion tube holder (top) and Dellorto holder (right). Sizing numbers arrowed.

found in the BMC/Rover range, will run best with the following emulsion tubes: F16/7772.7 for smaller engines (998cc-1098cc). F16/7772.7 or richer F2/7772.6 for larger engines (1275cc and up).

AIR CORRECTOR - SELECTION

The air corrector is used to tune top end performance but only over a very small range. Changes in air corrector size are made in increments of 10 at the very least and, more often, 20. Expect the range of air correctors to be from 150 to 230 for use with the main jets previously listed. Start with 160s and go leaner (a larger air corrector number). If the air corrector is too lean (large diameter) the engine will miss as it nears maximum rpm, and if too rich (too small a diameter) the engine will not produce optimum power. Within a small range of these two extremes will be the optimum size.

Weber auxiliary venturi. Position of sizing numbers arrowed.

The four 40mm Dellorto auxiliary venturis. Going left to right the 7848.4, 7848.1, 7848.2 and the 7848.3.

45 and 48 Dellorto auxiliary venturi with the position and sizing numbers arrowed.

For example, if 180 gives good power and going to a 200 causes the engine to misfire, while a 190 gives good power and no missing, the choice is either 180 or 190. Go with the 180 provided the power output remains the same. The Weber range of air correctors is 140, 145, 150, 155 and so on. The Dellorto range is the same except between 140 and 205 where they go up in .25 divisions (which is too fine and not really necessary).

Weber has it right here.

BMC/Rover Siamese intake port engines all start with 180 air correctors.

AUXILIARY VENTURI - SELECTION

The Weber number sequence is 3.5, 4.0, 4.5, 5.0 and so on, while the Dellorto is a suffix of .1, .2, .3 and .4. The numbers are stamped on the auxiliary venturi on the outer surface of the diameter near the entry of the air/fuel mixture passage way on Webers and cast in on the outside diameter on Dellortos. If the auxiliary venturi is not large enough the engine will falter at a certain point in the rpm range. What happens is that the engine just cannot get enough air/fuel mixture, so go up to the next size if this happens. The size of the auxiliary venturi has to be large enough but not larger than what is actually necessary. It is more a case of fitting the smallest auxiliary venturi possible compatible with top engine performance. Do not fit a 5.0 auxiliary venturi when a 4.5 is able to flow the necessary air/fuel mixture requirements. The table below gives the approximate numbers required on

40mm Dellorto auxiliary venturi and position of the number.

the basis of the ccs per cylinder of any engine -

Weber carburettors -
3.5 auxiliary venturi for up to 300cc per cylinder.
4.0 auxiliary venturi for up to 400cc per cylinder.
4.5 auxiliary venturi for up to 500cc per cylinder.
5.0 auxiliary venturi for 500cc per cylinder and up.

The three 45mm and 48mm Dellorto auxiliary venturis. The 8011.3 is on the left, the 8011.2 is in the centre and 8011.1 is on the right.

Weber accelerator pump jets. Position of size numbers arrowed.

Dellorto 40mm 7848.1 auxiliary venturi on the left, and 45mm - 48mm 8011.1 auxiliary venturi on the right.

Note: Weber uses a totally different shape of auxiliary venturi for the 40 DCOE as compared to the 45 and 48 DCOE. Be aware of this fact when ordering auxiliary venturis.

Dellorto DHLA 40mm carburettor auxiliary venturis are available in 4 types. They are: 7848.1, 7848.2, 7848.3 and 7848.4. 7848.1, 7848.2 and 7848.4 are very similar except that the 7848.2 has slotted standard length centre sections and 7848.4 has cut off short centre sections. The 7848.2 was for the Alfa Romeo DHLA 40H emission carburettors and 7848.4 is designed this way to reduce the depression in the centre section of the auxiliary venturi. These last two mentioned auxiliary venturis were for special applications.

7848.1 is the normal auxiliary venturi to use on all engines with individual cylinder capacities over 300cc. The 7848.3 auxiliary venturis can be used on engines which have individual cylinder capacities of less than 350cc or larger capacity standard engines up to 500cc per cylinder. The main difference between 7848.3 and the 7848.1 is the size of the passageway that the air/fuel mixture passes through to get to the centre section of the auxiliary venturi. The more air/fuel mixture going into the engine the larger the cross sectional area the passageway has to be. Use the 7848.3 if your engine responds well.

If the auxiliary venturis are too small, the top end engine performance will suddenly 'flatten off'. When this happens the ability of the auxiliary venturi to supply the required amount of air/fuel mixture has been exceeded. When this happens a change from 7848.3 to 7848.1 is required, but up until this point the 7848.3 will have slightly better low rpm response.

Weber has a slightly better range of auxiliary venturi feed slot sizes than Dellorto in this instance. Effectively, there are just two auxiliary venturi choices for 40mm DHLA carburettors: 7848.1 and 7848.3.

Dellorto 45mm and 48mm

Weber accelerator intake valve on the left has no discharge hole, centre left intake/discharge valve has a 0.5mm hole (listed as a 50), intake/discharge valve centre right has an 0.8mm hole (listed as an 80), intake/discharge valve on the right has a 1.0mm hole (listed as a 100).

Intake valve on the left has no discharge hole, centre intake/discharge valves are 50s and the valve on the right is a 100.

carburettors have had three auxiliary venturis available for them, in sizes 8011.1, 8011.2 and 8011.3. Of these, the 8011.1 is the most commonly used, the other two being specials for Lotus. The 8011.2s were for the DHLA45M carburettors fitted to a Lotus Turbo engine, while the 8011.4s were fitted to the tri-jet DHLA45D carburettors fitted to a Lotus engine.

BMC/Rover A-Series engines with Siamese inlet ports, for example, will use 4.5 coded auxiliary venturis in 40mm and 45mm DCOE Weber carburettors. The physical sizes and shapes of the auxiliary venturis are quite different between the two carburettors, so you must specify the type of carburettor when parts are ordered.

ACCELERATOR PUMP JET - SELECTION

The Weber range for average use is 30, 35, 40, 45, 50, 55 and 60 (the number referring to the hole size in hundredths of a millimeter). Dellorto have a wider range using the same principle of the given size being in hundredths of a millimeter, their range includes all of the Weber sizes and what can be arbitrarily termed half sizes. You can get 40, 41, 42, 43, 44, 45, and so on, from Dellorto which further assists fine tuning.

For all engines the accelerator pump jets have to be large enough to remove any trace of hesitation or stumbling when the accelerator pedal is depressed but not more than this. Find by trial and error the smallest pump jet that gives the best performance. Even with long duration camshafts it is possible to have smooth acceleration from quite low rpm in a high gear (full load) without any 'spitting back' or engine hesitation.

Overlarge accelerator pump jets will certainly cause excessive fuel use and to no advantage. A too small accelerator pump jet will cause the engine to die momentarily when the accelerator is depressed. Too large an accelerator pump jet causes the engine to 'bog' which effectively amounts to the same thing (a slow car) and then pick up after the excess mixture has been removed from the engine.

The following list gives a basic starting range of accelerator pump jet sizes for modified engines (ie with long-duration camshafts, worked cylinder heads, high compression, extractor exhaust systems). Try the smallest sizes first. The accelerator pump jets listed are average, so expect to come down on the recommended sizes for engines that are not overly modified (ie stock ports, stock compression, mild camshaft). There is a degree of overlapping with accelerator pump jets and sometimes a larger jet just has to be used. Dellorto have the edge here with their larger range.

Accelerator pump jet selection -

300cc per cylinder: 35s on the high side and 30s on the low side.
400cc per cylinder: 40s on the high side and 35s on the low side.
500cc per cylinder: 45s on the high side and 40s on the low side.
600cc per cylinder: 50s on the high side and 45s on the low side.
700cc per cylinder: 55s on the high side and 50s on the low side.
800cc per cylinder: 60s on the high side and 55s on the low side.
900cc per cylinder: 65s on the high side and 60s on the low side.

Note that with the Weber, the pump jet sizes go up in increments of 5 but they also use an accelerator pump discharge bleed to reduce the size of the shot. With this device there is adjustment to increase, or decrease, the fuel volume delivered. The intake and discharge valve is found in the bottom of the float chamber. This valve can have a large hole in it (maximum bleed-off) down to a small hole (minimal bleed-off) or no hole at all, in which case the full shot is delivered. Dellorto do not use this

type of arrangement and offer full shot fuel delivery all the time, however, their pump jets go up in increments of 1 which allows correct calibration within the confines of a full shot. Both systems work perfectly. When tuning a Weber, if, for instance, a 45 pump jet is too large and a 40 too small, fit the 45 pump jet and check the size of the bleed and increase the bleed size by 10 or 20, or more (larger hole to reduce the shot size) and lean the fuel shot off.

On BMC/Rover Siamese intake port engines expect the pump jets to range from 35 for the smaller 850cc engines to 45 on the larger 1275cc engines. It is most unlikely that a highly modified 1275 to 1330cc A-Series will ever need a larger accelerator pump jet than a 45.

ACCELERATOR PUMP INTAKE/DISCHARGE VALVE

If the accelerator pump intake valve has no hole in it, it is an accelerator pump intake valve and no fuel is bypassed back to the fuel chamber. If the accelerator pump intake valve has a hole in the side of it, it's an accelerator pump intake/discharge valve because fuel is discharged out of this hole when the accelerator pump is activated. The hole in the side of the intake/discharge valve is used to bleed off excess fuel or fuel not required to accelerate the engine cleanly. The sizing of the discharge hole is a tuning device used to set precisely the amount of fuel injected into the engine during accelerator pump action. It takes a bit of work to get this right, but get it right you can!

NEEDLE VALVE - SELECTION

The needle valve has to be large enough to keep up with the fuel demand of the engine; however, it should not be larger than is necessary. A large needle valve tends to gush the fuel in more than a small needle valve. If the needle valve is too large for the particular application, the valve will allow fresh fuel in very fast causing the float to rise and the needle valve to close but not fast enough to prevent the fuel level rising above the specified level and therefore giving a rich mixture. Sure, the level will drop again as the engine uses the fuel but what you have here is a constantly altering fuel level in the float chamber (up and down over the optimum level) and this is not desirable.

The following are recommended needle valve sizes for different engines.

One choke per cylinder applications -

Up to 300cc per cylinder use a 1.50.
Up to 400cc per cylinder use a 1.75.
Up to 500cc per cylinder use a 2.00.
Up to 600cc per cylinder use a 2.25.
Over 600cc per cylinder use a 2.50.

One choke per two cylinder applications -

Up to 250cc per cylinder use a 1.50.
Up to 300cc per cylinder use a 1.75.
Up to 350cc per cylinder use a 2.00.
Up to 400cc per cylinder use a 2.25.
Over 400cc per cylinder use a 2.50.

Chapter 5

Manifold preparation & carburettor fitting

INTAKE MANIFOLD - CHECKING & PREPARATION

Good progression, snappy mid-range performance and optimum top end power requires identical throttle opening in multiple carburettors from idle to full throttle. There is no substitute for absolute accuracy.

Intake manifold alignment is critical when linked throttle arms are used and this aspect of carburettor tuning is vital if maximum performance is required from engines with two or more carburettors. To take a common situation with twin sidedraughts on an in-line four-cylinder engine, the carburettor butterflies can be set perfectly at idle and just off idle yet, at half throttle and more, one set of butterflies can start to lag behind. At full throttle the situation is not usually so bad because the throttle spindle obscures the actual butterfly through its own physical size, so a bit of incorrect butterfly angle between the carburettors does not always affect

wide open throttle performance but the mid-range power will however be down in this situation. A common fault that causes butterflies to become unsynchronized at part throttle is poor intake stud to carburettor body alignment. This is a seriuos indictment of some manufacturers of intake manifolds who should know how important this feature is.

The importance of stud alignment

All intake manifolds must be checked for accuracy. There are various multiple carburettor linkages available but the most common is the linked throttle adjuster arrangement. However, this system will only be as good as the alignment of the intake manifold. If your manifold does not prove to be accurately made return it to the manufacturer for replacement or have the offending studs repositioned. The following simple test can be carried out anywhere using an

engineer's rule or a straight edge. If the manifold does not have any studs fitted just screw in some bolts. Lay the rule edge along the studs (see photo). If the studs are out you will be able to see it at a glance. The maximum allowable error is about 0.25mm (0.010in) but there should really be no perceptible alignment error.

If your manifold has two or three stud holes out by a reasonably small amount (1mm/0.040in) the easiest way of having a stud hole repositioned is to either 1) have the complete hole filled by TIG welding, reface the manifold surface and then re-drill and tap the hole in the correct position or 2) move the centre of the hole over by offset boring using a milling machine and fit a Helicoil which will correct the situation. Clearly the best thing to do is check the manifold properly before buying it and make sure you get a good one in the first place.

With the linked type of joining system the carburettors must be parallel (within 0.1mm/0.004in) to the intake manifold when O-ring gaskets are in place and the studs nuts are all tightened. They must also be perfectly in-line when looking straight down the throttle bores, or spindle to spindle, in both planes.

Another method of throttle action involves pushrods with balls and sockets at each end. This is quite common on six-cylinder in-line engines and comprises a sturdy, full length throttle bar with lever arms positioned directly above each carburettor's throttle arm. Once again component accuracy plays a large part in making sure all butterflies are operating in unison. The throttle bar lever arms must all be of equal pitch (within 0.1mm/0.004in). Otherwise if one arm has a shorter pitch, for instance, that particular carburettor will have advanced butterfly opening. The balls and sockets have to be kept in perfect condition if settings are to be reliably maintained. Manifold stud alignment is less critical when compared to the direct linkage system but, ideally, all intake manifolds should have accurate stud alignment.

Carburettor - checking fit
It is actually better to fit the carburettors to the manifold off the engine. With the integrity of the intake manifold checked (stud alignment and face flatness) fit all studs to the manifold and use Loctite (or any other suitable retaining compound) on the threads. Torque the studs to about 7 foot pounds using locked nuts. Check that all studs are at 90 degrees to the manifold face in all planes. This is to ensure that the carburettors will fit over the studs easily. All studs must protrude from the manifold's surface a minimum of 38mm (1.5in). The top ones should not be longer than 38mm but the bottom ones can be up

Use a straight edge to check that the studs are in alignment.

The typical twin-sidedraught manifold. Note how some of the carburettor mounting studs are angled and how, in the second picture, the carburettors (a Weber and a Dellorto) won't slide over the studs.

Simply excellent Maniflow made, steel inlet manifold for a single sidedraught carburettor application for Siamese intake port engines (A-Series). It doesn't get much better than this, and it's simple.

to 40mm (1.57in) as there is plenty of clearance. Fit a carburettor to each set of studs. It must fit easily and not bind on the studs all the way down to the manifold surface.

Anti-vibration mountings

Because they are susceptible to fuel frothing, Weber and Dellorto carburettors are installed with anti-vibration O-ring gaskets and related componentry. Kits comprise the large O-ring gasket (one per choke) and rubber grommets, metal caps and nylock nuts. In some applications

"Thackery" washers (short flat wound coil springs) are used beneath the carburettor securing nuts and these washers apply tension, via the carburettor body, to the O-rings.

Engines fitted with Webers or Dellortos as original equipment usually feature moulded rubber mounting blocks between the carburettor/s and the manifold. These blocks are excellent, but cannot usually be adapted to any other application.

Some of the plastic injection moulded O-ring carriers are not all that rigid, namely those designed to be

MISAB spacer (bottom right) and a O-ring carrier (bottom left). Stud, rubber grommet and a nylock nut assembled (top right). Stud, flat washer, thackery washer, flat washer and a nylock nut assembled (top left).

Right - Carburettor bolted to manifold showing the MISAB spacer in position. When the carburettor is finally bolted up, the distance between carburetor flange and the intake manifold flange must be equal in all positions to prevent air leakage. Note that the rubber grommets on the studs must not be completely compressed. Far right - Carburettor bolted to manifold showing O-ring and carrier in position. It often pays to use a small amount of silicon sealer to hold the O-ring in the plastic carrier to prevent the O-ring from falling out while fitting the carburettors. Note that Thackery washers must not be completely compressed.

used with 45 and 48mm carburettors. What happens is that the O-ring gets sucked out of the carrier because of the flimsy design of these larger carriers (compared to those of the 40mm carburettors). It's recommended that these larger plastic carriers are **not** used on any engine. Made in past years there have also been zinc based die-cast O-ring carriers which are excellent. The MISAB is the most readily available component these days and the best to use, however they do not last forever and should be replaced at the first sign of deterioration (cracking of the rubber).

CARBURETTOR/S - FITTING TO MANIFOLD

Ensure that throttle levers are fitted to the carburettor spindles and that the new lockwashers have their tabs bent over against their nuts. Place the O-rings over the studs and then fit the carburettors over the studs and settle them down onto the rubber O-rings. If twin sidedraughts are being fitted with a linked throttle bar arrangement,

MISAB spacer on the left, plastic 'O' ring carrier to suit a 45mm carburettor centre left, plastic 'O' ring carrier to suit a 40mm carburettor centre right and die-cast 'O' ring right.

interlink the throttle bars while the carburettors are on the bench and then carefully offer the manifold up to the sitting carburettors (not the other way around). If the carburettors are operated by an auxiliary bar then this is not a problem. Either way settle the carburettors on to O-rings, then fit the rubber grommets over the studs and then fit the metal grommet covers over the studs

Fitting the Nylock nuts is next and the top four nuts should be engaged first because there is the possibility of some interference with the carburettor body. All four nuts on the underside of the carburettors go on next and they will fit as there is plenty of clearance available. The carburettors may have to be lifted slightly to facilitate starting the nuts on the stud threads. Once all the nuts are wound on about three turns or so they will be clear of the carburettor body. The nuts are all wound down evenly until all are in contact with the metal grommet caps but the rubber grommet is still at its uncompressed height. In other words, no tension has been applied to the rubber grommet. Now wind down all nuts evenly, that is, one turn on each at a time until the metal grommet caps are within 1.5mm (0.059in) of the carburettor lugs. The O-rings will

now be compressed and the distance between the carburettor and the intake manifold will be approximately 5mm (0.196in). This clearance must be checked reasonably frequently if the rubbers are brand new. The aim is to tighten all the nuts so that a distance of 1.5mm between each grommet washer and the carburettor lug is achieved. If the rubber O-rings are not under

sufficient tension the carburettors may not have a good effective seal and any loose carburettor could run lean through the ensuing air leak.

To check the installation measure the distance between the carburettor flange and the intake manifold face at the top and bottom; the measurement must be equal (within 0.25mm/ 0.004in). If this is not the case expect possible leakage in the area with the largest measurement. The second reason for having the measurements equal is to ensure spindle alignment. It is vital to have the spindles exactly in line if a linked throttle arm system is used.

Irrespective of the type of throttle linkage used or the number of carburettors involved each carburettor must be mounted parallel to the manifold face. This is to preclude air leaks, which, if they exist, will interfere with the fuel/air mixture and general running of the engine. Check each

Spanish built DCO/SP 'competition type' sidedraught Weber carburettors have very little clearance around the stud hole. Makes fitting washers and nuts difficult.

carburettor frequently and especially if new rubbers have been fitted as they settle over time and what was quite tight can end up quite loose.

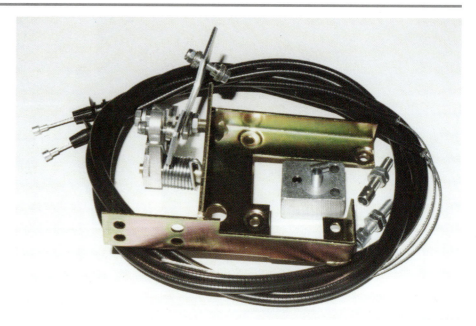

Single carburettor application twin cable throttle linkage.

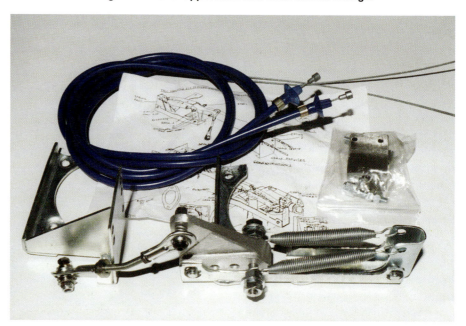

Twin carburettor application twin cable throttle linkage.

Chapter 6
Testing & set-up

What follows is a tuning sequence that can be applied to any Weber/Dellorto sidedraught carburettor installation irrespective of how many carburettors are involved. From this point on only tuning technique is discussed. Your carburettor/s should now be fitted with the chokes, venturis and jets chosen from the recommendations given in chapter 4 and, although alterations may well have to be made, the basic set-up should be nearly correct.

IDLE SPEED

The duration and overlap of the camshaft has a huge affect on the idling speed. The range of reasonable idling speeds for different camshaft configurations is as follows.

Standard camshaft (about 240 degrees duration, 30 degree overlap): 600-800rpm.

Mild performance camshaft (265 to 270 degrees duration, 50 degrees overlap): 600-1000rpm.

Medium performance camshaft (270 to 290 degrees duration, 70 degrees overlap): 100-1200rpm.

Racing camshaft (290 to 320 degrees duration, 90 degrees overlap): 1200-1500rpm.

FUEL LEVEL & NEEDLE VALVE OPERATION - CHECKING

Make sure float levels are correctly adjusted; the float level setting procedure is given in chapter 2.

If the engine has a mechanical fuel pump it should be cranked over or, if equipped with an electric fuel pump, the ignition turned on to allow the carburettor float bowls to fill with fuel. Check to see whether the needle valves are actually shutting off. The first indication of flooding will be fuel pouring out of the fuel bowl vent. This can happen with new carburettors but is more usual with used carburettors that have not been rebuilt correctly. Flooding can be caused by dirt or

debris in the needle valve body which prevents the needle from seating. With older, or badly rebuilt, carburettors the problem is more likely to be that the needle's tapered section is worn (grooved). Conversely, if the fuel bowls do not fill up, the needle could be jammed in the off position (dirt) and not allow the fuel to go into the float chamber. It is also possible for fuel line debris to block the mouth of the valve.

In the case of multiple carburettors, the next check is to remove the carburettor tops to see if the fuel levels of each carburettor look the same. Despite the fact that the floats are set to exacting measurements the ultimate aim of the procedure is to ensure that the fuel levels of each carburettor are the same and within optimal range on multiple carburettor set-ups. If necessary, set the floats of individual carburettors to give identical fuel levels, even if it means setting the floats to slightly non-standard settings.

THROTTLE - INITIAL ADJUSTMENT AND SYNCHRONIZATION

This involves the setting of each set of butterflies so that, no matter how many carburettors are fitted, all will be flowing the same amount of air at idle and throughout their entire operating range. The engine has to be running to synchronize the carburettors so the next thing that has to be done is get the engine running so that adjustments can be made.

Throttle arm fit

Note that each carburettor has a throttle arm of some description and this throttle arm fits on to the spindle at that point where two flats have been machined on it. The throttle arm has a precision fitting slot in it that allows the throttle arm to be a tight fit on to the spindle. If the throttle arm is a used one, check the fit of the throttle arm on to the spindle. The fit must be tight. If it isn't, the carburettor could have been used with the securing nut loose. This will mean that, although the nut and washer are tight, the throttle arm and spindle are not accurately located and could move. If this happens the carburettor will go out of synchronization. Replace any throttle arm that is not a good fit on the spindle. (An alternative to replacing the throttle arm is to braise the flats of the throttle arm slot and then, using needle files, file the slot out to fit the spindle. If the spindle is also worn this is often the only way of restoring the fit unless the spindle and throttle arm is to be replaced).

Throttle - initial setting (single carburettor)

On single carburettor set-ups, simply turn the throttle arm adjusting screw clockwise until you see the throttle arm just begin to move and then turn the

screw a quarter turn. If you are not sure of the throttle arm's state, turn the adjusting screw anti-clockwise and note if the arm moves. If the arm moves up then the throttle was cracked open. Turn the adjusting screw until the throttle arm stops moving. Slowly turn the adjusting screw clockwise until it contacts the throttle arm and then turn the adjusting screw a further quarter turn.

Throttle - initial set-up & synchronization (multiple carburettors)

For multiple carburettors linked by individual pushrods with socketed balls, undo the linkage arms on the main throttle bar. Check that the rods from socket center to socket center are equal in length (to within 0.5mm/0.020in). Turn each throttle arm adjusting screw anti-clockwise until there is no movement of the throttle arm and then turn each throttle arm adjusting screw clockwise until it just contacts the throttle arm. Now turn the throttle arm adjusting screw of each carburettor a quarter turn clockwise. Go to the main throttle arm and put light downward hand pressure on each throttle rod in turn and tighten to the securing nut. Do this to each throttle rod so that there is little or no play in the mechanism. The throttle synchronization will not be exactly right but close enough to start the engine.

For multiple carburettors using the more common linked throttle arm joining mechanism, turn the synchronizing screw on the joining link anti-clockwise until the end of the screw is clear of the central lever. Turn the adjusting screw on the throttle arm anti-clockwise until the arm does not move. Then turn the throttle arm adjusting screw clockwise until it just contacts the throttle arm. Now turn the synchronizing adjustment screw

clockwise until it contacts the central lever. Go back to the throttle arm adjusting screw and turn it a quarter turn. The throttle is now sufficiently synchronized to start the engine. At this point the throttle will work and all throttle arms will have a small amount of throttle wound on to effect an idle. The idle may not be smooth but it will be an idle.

IDLE MIXTURE - INITIAL ADJUSTMENT

Turn the idle mixture adjustment screws on Webers out 1 1/4 turns from the lightly seated fully in position. Turn the idle mixture adjustment screws on Dellortos out 3 1/2 turns from the lightly seated fully in position. The engine will start on this setting, but it may not run all that smoothly. When the engine has been warmed up and the idle speed screw set to give 1000rpm (or whatever rpm the engine will idle at) turn all of the idle mixture adjusting screws out a further 1/4 turn for Webers and a further 1/2 turn for Dellortos (that's a total of 1 1/2 turns for the Webers and 4 turns for the Dellortos). If the idle smoothness improves and the engine speed increases reset the idle speed back to 1000rpm (or whatever the original idle speed was).

Conversely, if the idle smoothness worsens and the idle speed rpm drops, turn the idle mixture adjustment screws back to what they were (1 1/4 for Webers and 3 1/2 for Dellortos). With engine basically stabilised then turn the idle mixture adjustments screws in a 1/4 turn for Webers and 1/2 a turn in for Dellortos (in total that's 1 full out turn for Webers and 3 full turns for Dellortos). The minimum turns out ever likely for a Weber carburettor are 3/4 to 7/8 and for a Dellorto it's 2 1/2 to 3 turns. The maximum number of turns out likely for a Weber is 1 1/2

Gunson synchronizing meter being used on a Dellorto carburettor.

The two main components of a linked throttle arm adjusting mechanism as used with multiple carburettors. One part of the arm fits to one carburettor's throttle spindle while the other piece of the arm fits to the throttle spindle of the adjacent carburettor.

and for a Dellorto is 5 1/2 turns.

If a Weber carburettor-equipped engine will only idle smoothly with the adjustment screws turned out around half a full turn from the lightly seated position, the chances are that the idle jet is too large. If the engine needs two full turns, or even more, the chances are that the idle jets are too small for the engine.

If a Dellorto carburettor-equipped engine will only idle smoothly with around 1 1/2 turns out from the fully seated position, the chances are that the idle jets are too large. If the engine needs 5 or 6 full turns out of the idle mixture adjusting screws, the chances are that the idle jets are too small. Further to this, if the engine hesitates when the throttle is opened slowly this is an indication of a weak mixture.

The idle mixture adjusting screws are turned individually with the engine being allowed to settle for 10 seconds between adjustments. Some idle mixture adjustment screws are going to affect the engine idle smoothness and rpm more than others. There is also

no absolute guarantee that all screws will end up being an identical number of turns out to achieve optimum idle smoothness. It's also quite likely that all of the idle mixture adjustment screws will end up being turned out an equal amount. Essentially, for the purposes of idling, the idle mixture adjusting screws need to be turned out the least number of part turns to effect the optimum idle smoothness.

If one idle screw makes no difference to the idle speed and smoothness of the engine there is something wrong with the idle circuit on that side of the carburettor. Something is blocked somewhere.

The object of this exercise is to get the engine idling as smoothly as possible, and at a reasonable speed. Idle speeds vary from engine to engine commensurate with the type of camshaft fitted. Suffice it to say that nothing can be done with any engine until it is idling smoothly (well, comparatively smoothly, anyway) and at a reasonable rpm relative to the engine's state of tune.

Gunson meter being used to synchronize a pair of carburettors. Here, the left-hand choke is being checked in the right-hand carburetor.

Gunson meter being used on a left-hand carburettor. Here, the right-hand choke of the left-hand carburettor is being checked. Use only the middle pair of chokes when synchronizing twin sidedraughts.

Just because an engine idles well with the idle jet and air bleed selection made from the listings given in this book, does not mean that the idle jet selection is correct. There's a bit more to it than that. The idle jet/air bleed combination is selected on the ability of the combination (idle jet fuel hole size and air bleed hole size) to provide the correct air/fuel mixture for the progression phase of the engine acceleration from idle. That is the air/fuel requirement for smooth engine acceleration the instant that the butterfly is moved from the idle stop, exposing the progression holes as they become downstream of the butterfly (as the butterfly sweeps over the progression holes, they come under engine vacuum) and air/fuel mixture is passed through the progression holes and into the engine. With the engine idling reasonably well, the next stage is throttle synchronisation.

Main throttle bar (arrowed) with pushrod connected. The bar must be strong and mounted in good bearings. Each of the actuating arms (arrowed) must be of the same length (bar center to ball).

A typical double-socket pushrod as commonly used on throttle bar systems. It is essential that the socket centre-to-socket centre dimensions of all pushrods are identical.

THROTTLE (LINKED THROTTLE ARM TYPE) - FINAL SYNCHRONIZATION

The linked throttle arm system is the most common for twin sidedraught installations. With the engine warmed-up and running at approximately 1200rpm, the two adjacent chokes of each carburettor pair are checked for air flow using a meter, of which there are several on the market. These meters give direct readings via a scale which is a part of the meter. This is the only way to go for a consistently accurate result.

On these linked throttle arm systems there will be only one screw for adjusting the idle speed (on the basis of moving the throttle arm away from the carburettor body) and this is on the left-hand carburettor (looking at the carburettors intakes). Other carburettors are linked to this arm via the synchronizing screw mechanism. The synchronizing screw is used to equalize the airflow of the adjacent carburettor/s to the left-hand carburettor. The left-hand carburettor is the lead carburettor and the other is adjusted to match it. The adjusting

screw of the right-hand carburettor can be adjusted to cause the right-hand carburettor to receive more or less air than the adjacent carburettor and cause engine rpm to rise or fall. Turning the adjusting screw clockwise increases the airflow into the right-hand carburettor and turning the adjusting screw anti-clockwise reduces the airflow. When the linkage is synchronized correctly, the throttle arm adjusting screw on the left-hand carburettor is used to adjust the idle speed of the engine.

Readings can be taken from any choke on either carburettor but make sure the readings are taken from the same choke each time because there are often slight variations from choke to choke in the same carburettor. Adjust the meter's setting so that the indicator is in the middle of the scale. Do not leave the meter over the carburettor choke for more than five seconds. Check each carburettor for flow and then adjust the second carburettor to equal the airflow of the left-hand 'datum' carburettor. If, for instance, the meter shows that the right-hand carburettor is flowing more

Far left - Throttle spindle in the idle position.
Left - Throttle spindle in the full-throttle position at which point the throttle butterflies must be fully open.

Main throttle bar at full throttle position. Note the downward angle of the main bar's throttle arms. Ideally, the arm's arc of travel is set so that the arc's center coincides with the horizontal plane.

air, turn the synchronizing screw of the adjacent carburettor anti-clockwise a small amount and take another pair of readings with the meter. Carry on adjusting the right-hand carburettor until the readings are equal. Flow meters are quite accurate and allow virtually perfect throttle synchronization to be achieved. Note: expect to have to check and re-adjust the synchronization every six months on a road car and before every meeting on a competition car.

THROTTLE (BAR & PUSHROD TYPE) - FINAL SYNCHRONIZATION

For multiple carburettors operated by a main throttle bar with arms and pushrods (balls and sockets at each end), the idle is set by undoing the clamps on the arms of the main throttle bar and adjusting each carburettor in turn. Each carburettor will have its own throttle adjustment screw and arm. With the engine fully-warmed and idling, place the meter over a choke on each adjacent carburettor and take a reading. Turn the screw(s) of the carburettor(s) that has the lowest reading on the scale and increase the flow of that carburettor(s) by turning the adjustment screw clockwise. When all carburettors are equal in air flow check the rev counter to see what the rpm is. If the revs are too high the rpm will have to be reduced by turning out each throttle adjusting screw an equal amount to cut the airflow down. Use the meter to check each carburettor so that when the engine is finally idling at the correct speed each carburettor is flowing exactly the same amount of air. The next stage is to synchronize the carburettors' throttles off idle.

It is essential to get an equal amount of air through each carburettor right up to and including full throttle. To achieve this the pushrods must each be identical in length (socket-to-socket) as must all the actuating arms of the throttle rod and all the actuating arms of the carburettor throttle spindles (from rod/spindle center to ball center). Throttle arms made by Weber or Dellorto will be correct.

The main throttle bar must be

"Motometer" being used. This type of flow meter fits over the end of the trumpet. The sight reading glass is integral with the meter.

sturdy and mounted firmly on to the engine so that there is no twist in the bar. If the bar is not strong enough the carburettors may not all receive equal movement. On six-cylinder in-line engines throttle bar flexing can be a problem. If the readily available arms which have a 7.9mm (0.312in) hole in them are used, the throttle bar must be made of high tensile steel and the throttle shaft must be well supported by at least four bearings (rose joints). This specification will eliminate the usual problems associated with long throttle shafts, arms and rods. The arc of travel of the throttle bar and arms must be equally dispersed about the horizontal plane to preclude any form of strain on the mechanism. The lengths of the adjustable rods can be altered to achieve this.

The carburettors can only be fully synchronized if all of the above requirements are met. The next stage is to slacken off the tensioning screws that clamp the arms to the throttle bar and, using light hand pressure, push down on the top socket. Maintain the

pressure on the socket and do up the screw on the throttle bar arm. Do this to all of the throttle arms. The reason for this is to position the mechanism so that there is no play in it anywhere and that any movement of the accelerator pedal results in equal carburettor throttle arm movement (equal air flow). To check this mechanically, wedge the accelerator pedal so that it gives a steady 2000 to 2500rpm and take meter readings from the same chokes that were used before to set the idle speed. The meter will have to be adjusted so that at the 2000 to 2500rpm the indicator is centrally situated in the scale. The Gunson unit, for example, has a vent adjustment that is twisted around to either reduce or increase the air flow and, in turn, raises or lowers the indicator on the scale. The Motometer has a screw adjustment that can be turned in or out to achieve a reduction or an increase in air flow. Do not hold the meter over a choke for more than five seconds.

If the readings are all identical the throttle bar and pushrod system is giving equal movement to the carburettor throttle arms: if not, the pushrods can be adjusted by lengthening or shortening them. The pushrods have right-hand and left-hand threads on their ends so turning the rod anti-clockwise shortens it and turning the rod clockwise increases the rod's length. This facility is now used to make minor adjustments to the length of the rods to give more or less carburettor throttle arm movement and to increase or decrease the air flow of a particular carburettor as required. Choose one carburettor as the standard, or datum, to which the other carburettors will be matched. Undo the locking nuts (slightly but not completely) and then turn the central rod anti-clockwise or clockwise. Check the airflow with the meter after each rod length alteration. Once the adjustments have been made and all carburettors are giving equal readings on the meter, turn the engine off and tighten all locknuts on the rods. While the engine is running during the tuning process keep an eye on its temperature. With all of the locking nuts tightened, start the engine and check all of the readings again just to make sure that none of the rods have moved which could have caused the settings to be lost.

At this point the idle has been set (equal meter readings on the chokes) and when the accelerator pedal is depressed the mechanism has been adjusted to give equal meter readings once again. The engine should be idling at 800 to1200rpm and have equal air flowing through each carburettor, which will not necessarily mean that the engine is idling smoothly.

If the engine will not accelerate in the free running state it certainly will not accelerate well under load. Also, just because the engine accelerates

Eurocarb offers this STE syncrometer (model SK).

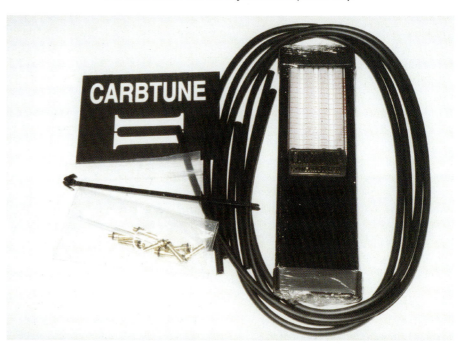

Eurocarb also offers this Carbtune which has four individual calibrated vacuum tubes.

smoothly under no load conditions, it does not mean it will accelerate under load (being driven down the road or tested on a rolling road).

At this point carburettors have the correct float levels and the throttles have been synchronized at idle and off idle. The idle mixture screws have been adjusted in or out to give the smoothest idle possible.

FULL THROTTLE CHECK

A very common problem with Webers and Dellortos is the fact that, in many instances, power is lost through the carburettors not actually achieving full throttle. As silly as it may sound, this is a very common error. **Always check** that the butterflies are at a full 90 degrees to the throttle bore when the accelerator pedal is pressed to the floor. To do this get someone to depress the accelerator pedal and using a torch look into the throttle bore and make sure that the butterflies are correctly angled. Failure to do so could result in a huge power loss ...

Check that the butterflies open fully on a regular basis, and certainly before racing begins. Throttle linkages being what they are (sometimes a bit flimsy) and the aggressive nature of racing, things can 'move' and full throttle can be lost.

Note that the full throttle stop on some linkage arms does not always stop the butterfly at the 90 degree point. What happens then is that the butterfly goes over centre and starts to restrict airflow. If this is happening, the throttle stop can be built up by brazing and then hand filed to give exactly 90 degrees of butterfly action.

If your engine goes better when the throttle is set less than fully open, the chances are that the chokes are too large for the particular engine. The choke is supposed to be the minimum restriction in the inlet tract, or the sizing factor for maximum air flow into the engine, and not the butterfly. Change the choke size to a smaller one and keep the butterfly in the straight, fully open position at full throttle.

IDLE JET ALTERATION (FUEL COMPONENT)

At this point the fuel jet can be altered if necessary. With the procedure followed to this point the carburettors

may well be synchronised, but the engine may not be idling decently at all. If the idle mixture screw turns out are too few, or too many, the idle jetting (air bleed component and or fuel component) is too large, or too small. If the idle jetting is grossly out, the time to re-evaluate the situation is now. The range of turns out with these carburettors is on average from between 7/8 of a turn to 1 1/2 turns for Webers, and from three full turns to 5 full turns for Dellortos.

If on Webers only a half a turn, or so, outward of the idle mixture adjustment screws is required to effect good idle, the chances are that the idle jet is too large. If, on the other hand, the idle mixture adjustment screws have to be turned out 2 1/2 turns to effect a good idle, the chances are that the idle jets are too small. Go up or down 5 increments (for example change from a 40 to 45).

If on Dellortos 1 1/2 turns, or so, is required to effect good engine idle, the chances are that the idle jets are too large. If on the other hand at least 6 1/2 turns, or more, is required to effect good engine idle, the chances are that the idle jets are too small. Change the jet size up or down by 5 increments.

If an individual idle mixture adjustment screw does not respond to adjustment the chances are that there is a blockage somewhere in the passageway. If blowing through with compressed air fails to remove the obstacle the lead plug may have to be removed. All manner of obstructions have been found in carburettors (including bits of lead plugs!). Check that the idle mixture adjusting screw's taper is not bent, making adjustment near impossible.

IGNITION TIMING - GENERAL

At this stage it must be pointed out

that your engine could be idling quite unevenly and be spitting back through the carburettors. This problem can be evident from all carburettor chokes consistently or at random or it may affect just one choke constantly. In the first instance, check the amount of ignition timing at idling speed with a strobe light. If the ignition timing advance is 2-8 degrees BTDC this may well be the sole cause of the spitting back through the carburettors. The problem is, in fact, nothing to do with the carburettors or their settings, though the carburettors are often wrongly blamed for this common fault. Before any further adjustments are made to the carburettors, the ignition timing must be checked.

Modified engines nearly always need a considerable amount of initial spark advance. If the engine being tuned does not have sufficient idle speed advance it will be impossible to get good clean performance below 2700 to 3500rpm (depending on the camshaft) without resorting to very large idle jets and lots of turns out on the idle screws, after which the richness of the mixture will tend to cover the lack of spark advance and reduce the spitting back but the engine will lack 'snap'. Low speed performance can be greatly improved, but the distributor advance mechanism will usually have to be altered to achieve this.

If the engine performs well everywhere except at or near maximum rpm (design strength maximum of the engine, that is) yet a misfire persists, check that the engine has enough spark advance. To test this, fit the carburettors with a small air corrector (160) and run the engine with two degrees more advance than before.

The following is an approximate guide to the maximum full advance degrees for a range of engines with

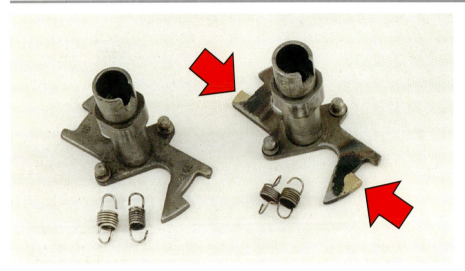

If the engine is set up with 15 or 16 degrees of advance at idle, the total advance will have to be limited. This is achieved by braising or mig welding the advance slot (arrowed right) which reduces the amount of travel. In this case the slot is reduced by 2mm so that the total advance does not exceed 35 or 36 degrees. Note the light springs which allow total advance at about 3300rpm.

various combustion chamber shapes (it is the combustion chamber shape that determines the optimum amount of total spark advance).

Bowl in piston: 32-36 degrees BTDC.
Hemi-head: 38-44 degrees BTDC.
Bathtub head: 32-36 degrees BTDC.
Pent roof (four valves per cylinder head): 30-32 degrees BTDC.
Wedge head: 28-38 degrees BTDC.

The correct amount of idle advance for any given engine is found by setting the idle speed to 1000 to 1200rpm with the idle speed timing to 10 degrees BTDC in the first instance and then increasing the ignition timing by two degrees (using a strobe light) and note the engine rpm. If the engine rpm does not increase, the advance is correct. If the rpm increased with the two degrees advance, advance the spark another two degrees to 16 degrees BTDC. Stop advancing the engine if the rpm does not increase. If the engine rpm increases to, say, 1700rpm by the time you have 16

degrees BTDC registering, reduce the idle speed back to 1200 via the carburettor idle screw.

Essentially, the final result should see the engine idling at the reasonable rate of about 1000 to 1200rpm with optimum spark advance so that it idles as smoothly as possible.

The downside of this increase in static advance is that, while the advance has been increased from, say, 8 degrees static advance to 16 degrees, the total advance has also gone up. So what used to be 8 degrees static and 38 degrees total advance (via the automatic advance mechanism), for example, suddenly becomes 16 degrees static and 46 degrees total advance. This must be altered to 16 degrees static and 38 degrees total advance. The amount of total advance required usually remains similar to standard even when an engine is modified.

This extra advance will have to be removed from the distributor because, if left, it could lead to poor top end performance and engine damage. The

solution is to reduce the mechanical advance mechanism's travel. This entails welding (tig welding or braising) the advance slot and remachining it. The slot will look the same but offer less travel. Checking and testing after welding is carried out using a strobe light. Basically, the advance added to the initial timing must be removed from the total timing. Note that some aftermarket distributors are adjustable and can be altered in five minutes, an example the being Mallory YL dual point.

For further details on ignition systems and their modifications there is another book in the SpeedPro Series called *How to Build and Power Tune Distributor Type Ignition Systems* which explains the situation much better than there is room for here.

IDLE JETS/AIR BLEEDS - FINAL SELECTION

The next stage is to check the idle mixture strength requirements for good progression. The idle mixture strength for idling purposes is not really in question because it is adjustable via the idle mixture adjusting screw. The mixture strength for progression has to be correct over a band, which means that the engine will run quite acceptably anywhere within this band. One side of the band will be towards the lean side and the other nearer the rich side. The objective of the fine tuning is to make sure that the mixture strength is in the middle of the band because, not only will the engine be as crisp as possible, it will give the maximum economy possible without any sacrifice in power.

Note that while it is quite possible to tune Weber and Dellorto carburettors very successfully without using scientific diagnostics, from here on Lambda and %CO readings are given for those intending to take their cars to a rolling road operator. The range of readings listed here are the general industry-wide ones.

The factor of turning the idle mixture adjusting screws to a position where it means that the engine has a very smooth idle, a light sound to the exhaust (at idle) and an acceptable idle speed Lambda, air/fuel or % CO scientific analysis reading (approximately 0.95-0.9 Lambda, 14:1.0-13.2:1 air/fuel, 1.6-3.3% CO) should, in the overall scheme of things, have very little to do with the immediate off idle performance, or the 'progression phase' as it is termed, of the engine.

The immediate off idle response of an engine equipped with Webers or Dellortos is controlled by the idle jets, although the accelerator pumps do have some influence. If the engine is accelerated slowly the accelerator pumps have no effect, so checking the idle jets on the basis of richness/leanness is based on this factor: the engine is accelerated slowly, but there must be no hesitation at all. One thing is for sure, if the idle jets are too small, or too large, an engine will not accelerate cleanly at all.

What happens in these carburettors is that the idle jets feed the idle mixture adjusting screws which, in turn, allow the engine its only source of air/fuel mixture when the butterfly is set to a suitable idle speed (choke not operating, of course). When the throttle is opened the butterfly sweeps past the progression holes which then become downstream of the edge of the butterfly and, therefore, subject to engine vacuum. In this situation air/fuel mixture from the idle jets is being admitted to the engine via the idle mixture adjusting screws and the progression holes. So, effectively, the idle jets are sized to the air/fuel requirements of the engine during the progression phase and not the requirements of the engine at idle as the name of the jets involved might suggest. More important is the fact that the idle mixture requirement of an engine is adjustable via the idle mixture adjusting screws, whereas the amount of 'progression phase' air/fuel mixture is set by the idle jetting and is only adjustable by changing the jet sizes (air and fuel components).

The air/fuel mixture which is supplied to the idle mixture adjusting screws and the progression holes is mixed by the idle jet and then travels to the progression holes and on to the idle mixture adjusting screws in same passageway. This is how closely these items are linked to each other.

A noteworthy point is, of course, that the adjustment of the idle mixture adjusting screws can, to a small degree, be used to advantage for tuning purposes on the basis of richening the 'progression phase' mixture, but at the expense of the idle speed Lambda, air/fuel or % CO content and the absolute smoothness of the idle.

Take, for example, an engine that idles perfectly well when there is, say, 1 1/4 turns out of each idle mixture adjustment screw (Weber carburettors in this example) and idles well, and can be slowly accelerated cleanly when not under load. All seems perfect, yet, when that engine is subjected to load (as in the car being driven on the road or track), the engine hesitates. The problem here is one of 'progression phase' mixture weakness. The first thing to do is to see if the hesitation can be removed (without changing the idle jetting) by increasing the amount of air/fuel mixture that is constantly being admitted to the engine via the idle mixture adjusting screws. This means increasing the idle screw turns out a further 1/4 turn on a Weber and a 1/2 turn on a Dellorto. This

will upset the idle 0.95-0.9 Lambda, 14.0:1-13.2:1 air/fuel or 1.6-3.3% CO readings and, perhaps, the idle smoothness, but, often, the effect on these two factors is minimal and the 'progression phase' hesitation disappears. What has happened in such cases is that little bit of extra idle mixture provided is enough to supply the engine with just enough air/fuel mixture to prevent the hesitation.

If the idle mixture Lambda, air/fuel, % CO setting goes too high then the size of the idle jetting needs to be altered, this may be as simple as reducing the size of the air bleed hole on a Weber or increasing the fuel jet size by 2 on a Dellorto. It frequently does not take much of an adjustment to remove 'acceleration phase' hesitation under load. A large adjustment which causes a considerable increase in the air/fuel richness may well remove the 'acceleration phase' hesitation, but the 'progression phase' air/fuel may now be very rich, which is wasteful of fuel and quite unnecessary. The air/fuel mixture needs to be rich enough, but no more. If every component of the jetting of a Weber or Dellorto is over-rich to cover problems, the result is exceedingly poor fuel economy. Whether the car is used on the road or the race track, it might as well be adjusted correctly for best possible power and within reason, economy. On a road car equipped with Webers or Dellortos there is no joy in having to fill up with fuel at every second gas station/garage. This can become very wearing but, unless the time is taken to narrow down the range of possible adjustments, is often the result.

The use of an exhaust gas analyser which gives a direct reading can be of immense help in the sorting out of air/fuel mixture ratios. These often inexpensive machines can be hooked up to the engine's exhaust system and give you an exact reading within five minutes and remove all possible doubt as to whether the engine isrunning too rich or lean. On average the idle mixture readings will between 0.95-0.9 Lambda, 14.0:1-13.2:1 air/fuel or 1.6-3.3% CO and no engine should go over (rich) 0.95 Lambda, 13.2 air/fuel or 3.3% CO as the ideal and obtainable setting for a road going engine.

Weber

In the first instance the idle mixture adjusting screws are set at the number of turns out which gives a 'good idle.' Then, with the engine idling as smoothly as possible, very slowly increase the engine speed. The object of the exercise is to check if the mixture strength is sufficient to match the slowly opening throttle without there being any hesitation/missing/stumble amnd without any assistance from the accelerator pump fuel delivery. If the engine stumbles/hesitates, the mixture is too weak. If the idle mixture adjusting screws are set to, say, 1 1/4 turns to effect good idle, but the engine stumbles when slowly accelerated, turn the idle mixture adjusting to 1 3/4 of a turn out and try it again. If the engine still stumbles/hesitates, the jetting will have to be altered as it is unlikely that 'progression phase' acceleration is going to improve without the idle speed readings going beyond 0.95 Lambda, 13.2:1 air/fuel or 3.3% CO. So, if the engine stumbles/hesitates, go to a richer combination (from, say, F2 to F11 but with the same prefix numbers - a jump richer by two). If the engine still stumbles, go to F9 and, finally, to F6. The idle mixture adjusting screws will need to be re-adjusted after any jet changing.

If none of these air bleed reductions cure the problem go up in prefix numbers, for example, from 45 to 50 and start again with an F2 and work down in two stage increments (F2, F11, F9, F8, F6). If, for example, the jetting fitted to the engine was started at 45F2 and after proceeding down the range from there to 45F11, 45F9, 45F8 and, finally, 45F6 and the stumble is still present, change the fuel jet number to 50F2 and proceed to go through the range again 50F11, 50F9, 50F8 and 50F6. Repeat this procedure until a suitable combination is found.

The converse is also true that if the idle jet selected is too rich and there is no hesitation the mixture can be leaned off until the engine does hesitate. The mixture strength can then be made richer by jetting up until the hesitation ceases. An over-rich mixture does not give top performance.

The idle screws have remained untouched so far during this test. With a jet found that gives no hesitation, turn all idle mixture screws in a quarter turn and note any difference to the idle (smoother) then turn them back to where they were before and then turn them out a quarter turn and note any differences. The smoother the engine runs the better. All screws must be set to the same number of turns and part turns out. Remember to allow ten seconds once the last screw has been turned for the engine to settle and stabilize using the new mixture. Find the best position for the idle mixture adjusting screw for the smoothest idle possible (least exhaust pulsing felt by hand at the end of the exhaust pipe).

By leaving the idle mixture screws in a set position and changing the jetting to suit, the richness/leanness factor is narrowed right down to the equivalent of about a quarter turn of the idle mixture screws; ideal for idling purposes and offering good progression with the engine in the free revving state.

Under load, the setting may prove to be not quite rich enough. Just because the engine accelerates smoothly under no load conditions it does not mean it will accelerate cleanly under load (being driven down the road or tested on a rolling road). If this happens move richer again until the hesitation stops under load conditions. The best test for a road vehicle is on an inclined section of road (not too steep) using top gear. If the engine mixture strength is too weak this sort of test will show it up. Jet up enough, but only enough, to provide for good progression. High speed is not necessary for this testing, rather low end pulling power.

If the idle jet (prefix number) is changed to overcome hesitation under load, the idle mixture will be richer and the idle mixture adjusting screws will have to be altered (turned in). This is exactly what they are designed for; turn them in a half turn or whatever is necessary to restore idle smoothness or remove the 'heaviness' that a too rich mixture gives.

Dellorto

Although Dellorto carburettors have a range of 10 idle jet holders, only the 7850.1, 7850.3, 7850.6, 7850.7, 7850.2 and 7850.8 (lean to rich) need to be taken into consideration for performance applications. The 8750.1 idle jet holder is an ideal starting point.

Dellorto carburettors can be a little bit less complicated than Webers when it comes to selecting the correct idle jet. The reason for this is the reasonably widespread practice of fitting either 7850.2, 7850.6 or 7850.1 idle jet holders to all engines initially, and increasing the size of the idle jet until acceptable idle speed running and smooth progression phase performance are obtained. This means that only the idle jet is changed to effect mixture corrections. This does not always prove to give optimum mixture strength, but frequently is very close; so close as to be acceptable. As with the Weber, what matters with idle jet selection is how well the engine accelerates from idle speed through the 'progression phase'.

What happens is that the air/fuel mixture that goes to the progression holes and the idle adjustment screw is supplied by the same jet: the idle jet and its air bleed component. The major difference in dealing with the two factors is that the idle mixture is adjustable via the idle mixture adjustment screws. The progression phase air/fuel mixture is not adjustable, other than by changing jets. Invariably, if the jetting is suitable for the progression phase of the engine, the idle mixture adjusting screws can be adjusted to suit and allow the engine to idle perfectly. If the idle jets are not correct for the acceleration phase, they may well be suitable for perfect idle. So, it's quite possible to fit jets to an engine and get it idling perfectly by adjusting the mixture screws, only to find that once the throttle is opened the engine will not go through the 'progression phase' because the jets are too lean.

If this happens, the idle jet size (fuel component) must be increased, such as going from a 45 to 50. With the Dellorto, it's possible to increase, or decrease for that matter, the jetting by 1 or 2 sizes at a time until the richness factor is correct. If the jetting is suspected to be too rich for the progression phase, reduce the jet size by 1 or 2 sizes and try again. Adjust the idle mixture adjusting screw at each jet change just to make sure that the idle mixture is actually at 2% to 2.5% CO (an acceptable air/fuel mixture).

For Dellortos the range of air bleed and fuel jet combinations is vast,

and five times that of Weber. In the interests of cost and overall efficiency, however, the range can be narrowed down by using just five idle jet holders (lean to rich): 7850.1, 7850.6, 7850.7, 7850.2 and 7850.8. The 7850.3 size falls between the 7850.6 and the 7850.2, and isn't really worth having over the other two.

The first step is to fit the fuel jet suggested in chapter 4 to a 7850.1, 7850.6 or 7850.2 idle jet holder and, with the idle mixture adjusting screws set to the number of turns that effects best idle, check the 'progression phase' by very slowly increasing the engine rpm (no higher than 2500rpm).

If the engine stumbles/hesitates, turn the idle mixture adjusting screws out a further 1/2 turn and try again. If the engine still stumbles/hesitates, turn the idle mixture adjusting screws out to a maximum of 5 1/2 full turns from the lightly seated position. If the engine still stumbles/hesitates, the fuel jet size will have to be increased until the hesitation disappears. A jet size change will mean resetting the idle mixture adjustment screws to give the smoothest idle. Using an exhaust gas analyser is recommended in the interests of getting the air/fuel mixture right at idle and through the 'progression phase.'

The 7850.1 idle jet holder has a 1.4mm diameter hole in the side of the idle jet holder and an axial bore diameter of 3.0mm. Any engine that has 400cc per cylinder and above will usually use this idle jet holder. The amount of air bleed from a 7850.1 is actually very similar to an F2 from the Weber range and the F2 is only a middle of the range air bleed (when applied to richness and leanness capabilities with a given fuel jet).

If there is no hesitation with the recommended idle jet leave the jetting just as it is and see how the

engine performs under road or track conditions. The fuel jet size may prove to be too large but this is not likely. The idle jets recommended tend toward leanness so, if anything, will require increasing in jet size to overcome obvious leanness (hesitation during the progression phase).

If there is hesitation, increase the idle jet size by 2 or 3 and test again; continue to increase the jet size as necessary but not beyond reasonable limits (plus 10 or 12 above the originally recommended size). If necessary change the idle jet holder to a 7850.6 and increase jet sizes from the recommended starting size up to plus 10 or 12.

Continue to change the fuel jet sizes by going up in 2s or 3s over the range suggested with each idle jet holder. The first combination that gives no acceleration phase hesitation will be very near and road or rolling road testing will narrow the choice down to the correct idle jet holder and fuel jet.

There is nothing like actual load conditions to show up mixture weakness. The fitting of larger jets (one or two sizes up) may be necessary to eliminate this weakness and ensure the engine is supplied with the correct mixture strength, which will result in best performance anywhere in the rpm range under any load condition.

The 7850.1 is a middle of the range air bleed from Dellorto and all other idle jet holders are going to supply a richer mixture strength in sequence for a given fuel jet size.

The method of tuning the Dellorto varies from that for the Weber because the idle jet and the air bleed is separate on the Dellorto. Weber has the air bleed sizes altered through the range while the fuel component stays the same. The Dellorto has the jet sizes altered while the air bleed stays the same; the end result is the same.

IDLE MIXTURE SCREWS - FINAL SETTING

Characteristically, on non 'idle by-pass' Weber carburettors, the idle adjustment screws (0.7mm x 5mm diameter thread pitch) will be turned out 1 to 1 1/2 turns from the fully seated position. For non 'idle by-pass' Dellorto carburettors, with their 0.4mm pitch x 7mm diameter fine threaded adjustment screws, it will be 3 1/2 to 6 1/2 turns (half this for the Dellorto coarse thread idle adjustment screw carburettors). There have been two idle adjustment screw thread pitches used on Dellorto sidedraught carburettors. Those enclosed in towers are all very fine thread, while those not in towers are similar to the Weber. Weber always used the same thread pitch for idle adjustment screws.

Dellorto used a fine thread to make the idle adjustment as fine as possible (unnecessarily fine in the opinion of many). A complication with this, however, revolves around the fact that it's not a particularly brilliant idea to use a fine thread in aluminium. Failure of the Dellorto idle adjustment screws is common, and the carburettor is not really serviceable once the idle adjustment screw slot has been damaged by excessive turning pressure due to a jammed/seized adjustment screw. All Dellorto carburettors which are equipped with idle screw adjustment towers should have plastic caps fitted over them to stop dirt and grime getting down into the threads, and they should be lubricated with CRC or WD40 occasionally.

The objective is to set the idle screws so that the engine sounds 'light' as opposed to 'heavy' (heavy indicating richness). The leaner the better for emissions but, obviously, the engine has to be running smoothly. It is recommended that an exhaust gas analyser be used just to be sure

Dellorto accelerator pump actuating rod and arm showing the two nuts (arrowed) that are used to adjust the position of the arm which contacts the accelerator pump diaphragm. Avoid touching these nuts unless there is clearly an adjustment problem.

of what the Lambda, air/fuel or % CO content of the exhaust gas is, but failing this continue with the following procedure.

With the static ignition advance determined and set, the progression phase checked and the idle jets optimized, the idle screws can now be adjusted so that the engine has the smoothest possible idle.

With the engine warm and running at a suitable idle speed, screw all the mixture adjusting screws in a half turn and note the difference to the engine's idling speed and smoothness. If the idle becomes smoother and the rpm increases the engine has responded to the leaner mixture. Continue to turn all of the idle mixture screws in (a quarter turn at a time) until the idle perceptibly roughens. At this point the mixture has gone too lean. Turn the screws back to the position that produced the smoothest idle so far (make a note of how many turns out this position represents). Now turn all the screws out (a half turn at a

time) from that position until the idle becomes 'heavy.' Work out the middle position between over-weakness and over-richness and set all the screws to that position.

It is fairly usual for the optimum setting to be the same for each mixture screw but this is not always the case, so do the following test. Go to the end of the exhaust pipe and place the palm of your hand about 50mm/2in away from it and feel the individual exhaust pulses against your hand. If the engine has a wild camshaft, the engine may idle roughly but it should still idle uniformly. If one cylinder is a bit lean, that cylinder will miss occasionally and this will be noticeable as a missed exhaust pulse against your hand; you may also be able to see 'spit back' from the carburettor/s.

To locate the offending cylinder, start by turning the idle mixture screw of the choke nearest the front of the engine one full turn from its current position. Then go to the exhaust pipe and feel the exhaust pulses. If the miss has gone the cylinder (or cylinders) fed by this choke is the offending one. If the exhaust pulse is unchanged, turn the idle mixture screw back to its previous position and move to the next idle screw and repeat the procedure. Once the offending cylinder (or cylinders) has been located, set that idle mixture screw out a sufficient number of turns or part turns to stop the miss, but no more. Obviously this particular cylinder is always going to have an odd number of turns out compared to the other cylinders.

The ideal position for each idle screw is between when the mixture strength weakens perceptively (engine misses) and when it turns rich (idle sounds 'heavy'), or that point that gives the highest idle speed with the maximum idle smoothness.

When engine idle is at its smoothest via the idle mixture adjustment, if necessary, reset the idle rpm by adjusting the throttle arm (slow the engine down).

Once the engine is warm it should start at the turn of the key without any depressing of the throttle to activate the accelerator pumps. If the engine is a bit reluctant to start, depress the accelerator pedal quickly to quarter distance to activate the accelerator pumps and the engine should start. Cold engines will usually respond to one full pump of the accelerator peda.

'IDLE BY-PASS' CIRCUITRY CARBURETTORS

There are Weber and Dellorto carburettors which have been fitted with 'idle by-pass' circuitry and Weber and Dellorto sidedraught carburettors that have not been fitted with it. In the cases where 'idle by-pass' circuitry has been fitted virtually all of them can be made in-operative by turning the adjustment screws fully in. The exception to this are some emission

type 40mm Weber carburettors which cannot be adjusted, the amount of 'idle by-pass' is factory set via drilled holes.

The majority of emission type 40mm sidedraught Webers all had adjustable 'idle by-pass' circuitry fitted (with recessed adjustment screws). The later Spanish built 40mm DCOE 151 and 45mm DCOE 152 carburettors also have idle by-pass' circuitry fitted to them, with the adjustment screws in cast in towers on each side of the body. A plastic cap covers each adjustment screw and locking nut. That's the same principle of operation for both types of adjustable carburettor but a different method of achieving it.

When Dellorto started making their sidedraught carburettors in the late 1960s, none of their carburettors were fitted with 'idle by-pass' circuitry but in the mid-1970s they changed this and all subsequent Dellorto sidedraught carburettors came equipped with 'idle by-pass' circuitry. The 'universal performance' Dellortos have external adjustment 'idle by-pass' screws and locking nuts while the emission controlled carburettors have recessed 'idle by-pass' adjustment screws.

Consider the 'idle by-pass' circuitry to be a tuning tool with a few functions. One function is to enable the idle mixture air flow of each individual choke to be leaned off quickly and easily. This means that all chokes, no matter how many carburettors are on an engine, can be set to have minimum and equal emissions (equal air flow). If a carburettor has a slightly bent spindle or a poor fitting butterfly, for example, the air content for each cylinder can still be adjusted via this means, whereas normally it wouldn't be able to be adjusted at all. Having an 'idle by-pass' system on these carburettors virtually guarantees that the minimum idle exhaust emissions

possible from the combination can be achieved. Having 'idle by-pass' circuitry also means that the butterflies can be set in the virtually shut off position at idle, so allowing the edges of the butterflies to 'sweep' over the progression holes from this position. This gives the maximum possible progression phase 'sweep' of the top edge of the butterflies past the progression holes as the throttle is opened. This causes the maximum amount of air/fuel mixture to be drawn through the progression holes and into the engine. This is all achieved within the confines of the idle jet size of course which control the richness/leanness factor of the air/fuel mixture and the number and size of the progression holes in the particular model of carburettor being used.

On Weber and Dellorto carburettors which have 'idle by-pass' circuitry the principle of operation is virtually identical. What all of these systems do is route air only passed the butterfly when the butterfly is in the closed position as at idle. The normal air/fuel circuit that is common to all sidedraught Weber and Dellorto carburettors is still working as per usual (this is the source of the fuel used at idle) but the 'idle by-pass' circuit, if in operation, 'dilutes' that air/fuel mixture.

The instant the throttle is opened, the by-pass circuitry is in-operative as there is no longer any vacuum differential from one side of the butterfly and the other. The progression phase off idle is then much the same as any other model of sidedraught Weber or Dellorto carburettor.

When 'idle by-pass' equipped carburettors of either type are being used on competition engines, most mechanics and engine builders just screw the 'idle by-pass' adjustment screws fully in so that the circuits are

This Weber has a plastic cap (arrowed), over the tower mounted 'idle by-pass' adjustment screw and locking nut.

inoperative and use an amount of butterfly opening to achieve idle and adjust the idle mixture adjustment screws to allow sufficient air/fuel mixture to enter the engine to effect idle. The air that allows the engine to idle is being drawn into the engine from around the edges of the nearly shut off butterflies and through the idle circuit. The fuel content to run the engine coming solely from the idle circuit and through the adjustment screw and is an air/fuel emulsion after being mixed at the idle jet.

Note: Being able to adjust the butterflies back to the near shut off position, and have them 'sweep' fully over all of the progression holes, doesn't alter the fact that a carburettor that has a lot of (four or five) large diameter progression holes is still going to have a technically weaker progression mixture than a carburettor which has two or three small diameter progression holes (maximum possible progression phase richness).

Adjusting idle by-pass systems

In the first instance close off the 'idle by-pass' adjustment screws and get the engine to idle at a suitable speed using an amount of throttle and get the engine to idle smoothly using the idle

mixture adjustment screws. This is all you can do for example on an engine with carburettors which don't have 'idle by-pass' circuitry anyway and up to this point this is the 'normal' method of tuning these carburettors. Then test the engine for progression phase or off idle acceleration. If the engine accelerates well without any hesitation then the idle jet or progression jet is correct for the particular engine. Base line engine performance is established when you do this.

If you now connect the engine to scientific diagnostic analaysis equipment the reading could be anything from 0.90-0.82 Lambda, 13.2:1-12.2:1 air/fuel or 3.3-5.5% CO or even a bit more. This is really quite high, and totally unnecessary on a road going car. Even a racing engine will not require such a high reading. With the engine idling, check the air flow through the engine with a meter and find the highest flowing cylinder. Now open the 'idle by-pass' screw of the cylinder flowing the least air and open it until the air flow equals the highest flowing cylinder. Adjust the other cylinders so that they are all equal.

The 'universal performance' type 40mm, 45mm and 48mm DHLA Dellorto has externally situated 'idle by-pass' screws and locking nut (arrowed).

On this Dellorto carburettor the progression hole plug (A) has a Philips head, and the idle by-pass adjustment screw is in a tower (B).

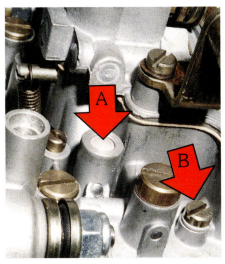

On this Dellorto, the idle by-pass tower is plugged (A), and the choke location screw is shown at B.

Adjust the idle speed down if the engine turns over too fast, and then check that each cylinder is flowing the exact same amount of air. It doesn't take too much effort to do this and the idle will be as good as it's possible to get it, and the CO and the HC the lowest possible.

Many mechanics don't bother to go to this trouble for racing engines as they claim it's unnecessary. However, racing engines do foul spark plugs from time to time, but not if the idle mixture has been adjusted correctly as suggested. If the carburettors are equipped with 'idle by-pass' circuitry, it's advisable to use it for this purpose. Not only will the idle air be equal, but the butterflies can be set in the near shut off position for near maximum butterfly sweep past the progression holes.

The limit to the adjustment of the 'idle by-pass' system is when the butterflies are in the near shut off position; the individual chokes have equal air going through them, the idle speed of the engine is as required and the engine is able to accelerate from the off idle position as well as it did when the 'idle by-pass' system was in-operative. Too much 'idle by-pass' adjustment air going into the engine can upset progression phase.

It's normal to end up with a very small amount of main throttle butterfly opening and an amount of 'idle by-pass' and a very low idle mixture reading such as 0.95-0.92 Lambda, 14.0:1-13.5:1 air/fuel or 1.6-2.5% CO and have the same acceleration phase performance that you had before.

CO readings and air'fuel ratio readings

These two factors are quite easy to co-relate on an approximate basis and a level of understanding is required. Most rolling roads use Lambda or % CO exhaust gas analysis equipment which uses a probe fitted into the exhaust system to give a digital reading for %

CO or a printout for Lambda. While air/fuel ratios are what are effectively being discussed, they are seldom used as meters these days. They were used on earlier times but they were slow reacting. Lambda, on the other hand, is instant.

The chemically correct air/fuel for optimum and best possible burning (clean burning and low emissions) is 14.7:1 or Lambda 1.00 or 0.5% CO (carbon monoxide). All cars made today have to comply with USA and EU emissions standards and the fuel injection systems used are now 'focused' to run the engine constantly at 14.7:1 air/fuel or Lambda 1.00 or 0.5% CO on a constant basis. The approximate equivalent Lambda to % CO to air/fuel ratios are listed in the accompanying chart.

Consider a suitable idle mixture to be 0.95-0.93 Lambda, 14.0:1 to 13.7:1 air/fuel or 1.6-2.0% CO for engines fitted with carburettors with 'idle by-pass' circuitry. All other engines, including racing engines, should be able to idle with a Lambda of 0.93-0.90, 13.7-13.2 air/fuel ratio or 2.0-3.3% CO idle mixture. Expect a full power mixture to be dependant on the engine design, but expect this to be between 0.82-0.88 Lambda, 13.0:1 to 12.2:1 air/fuel ratio or 3.8-5.9% CO.

Note that the 'California Air Resource Board' has been driving the world's auto manufacturers along the path of cleaner burning engines since about 1970. While working within the realms of the technology of the day they have never the less been relentlessly getting the various engine manufacturers to reduce emissions.

ACCELERATOR PUMP JETS - FINAL SELECTION

The fixed jetting of the carburettor cannot respond quickly when the throttle is opened rapidly. The

Lambda	Air/Fuel	% CO
0.80	11.8:1	8.0
0.81	11.9:1	7.3
0.82	12.0:1	6.5
0.83	12.2:1	5.9
0.84	12.4:1	5.4
0.85	12.5:1	5.0
0.86	12.6:1	4.85
0.87	12.8:1	4.35
0.88	13.0:1	3.8
0.90	13.2:1	3.3
0.91	13.4:1	2.85
0.92	13.5:1	2.6
0.93	13.7:1	2.15
0.94	13.8:1	1.9
0.95	14.0:1	1.6
0.96	14.1:1	1.4
0.97	14.3:1	1.0
0.98	14.4:1	0.8
0.99	14.6:1	0.6
1.00	14.7:1	0.5
1.01	14.8:1	0.6
1.02	15.0:1	0.3
1.03	15.1:1	0.15
1.04	15.2:1	0.2
1.05	15.4:1	0.15

Lambda, air/fuel, % CO conversion chart.

accelerator pump supplies fuel immediately and in sufficient volume to match the amount of air available to the engine when the throttle is opened quickly. Because the fuel injected by the accelerator pumps is not emulsified, more than is strictly necessary is injected into the air stream and much of it is passed out of the engine unburnt.

Once the main circuit comes into operation and the auxiliary venturi is flowing air/fuel mixture, the accelerator pump is no longer needed. If the engine emits a lot of black smoke when the throttle is suddenly opened this is a sign that the accelerator pump jets are too large. However, the next size down in pump jets (for Weber) may well cause the engine to hesitate. On Dellortos the larger range of jet sizes allows for very fine adjustment.

Fit the accelerator pump jet that is the smallest possible compatible with the engine having good acceleration without any hesitation. Try the recommended pump jet and increase the jet size if it proves to be insufficient. Conversely, try the next size down from the recommended size just to see if the engine can operate with that size. If the test proves successful, change down a size.

If the engine hesitates irrespective of the pump jet size installed in the carburettor, check the timing of the fuel delivery from the accelerator pump discharge nozzles compared to the throttle movement. This action has to be simultaneous.

On Dellortos check that the arm that operates the diaphragm is in fact in contact with the diaphragm base. These arms are factory set and usually correct but if the carburettor is second-hand they may have been altered. If during quick throttle movement no fuel is ejected, the arm may well be moving but not moving the diaphragm. The two locking nuts will have to be re-adjusted to take out the play

that exists so that the diaphragm is activated immediately the throttle arm is moved. Note that it is possible to set the accelerator pump arm too high. If the nuts are wound up too far on the available thread, the arm can jam. If there is more than 0.3in/8mm of thread showing under the nut, it's undoubtedly too much.

When Weber or Dellorto carburettors are full of fuel they will squirt fuel well over 3 feet/1 meter when the throttle is opened quickly. If there is any doubt about accelerator pump effectiveness, remove the carburettors from the engine (making sure that no fuel is spilled from them) and set them up level on a bench. To check the pump action depress the throttle about 20 per cent very quickly. Each nozzle should squirt fuel out of the choke and onto the bench and there should be a puddle of fuel in the throttle bore. There is a limited number of times that this can be done as the float chamber will empty itself. Pump action has to be instantaneous for correct engine acceleration. The quantity of fuel is correct when one full throttle arm depression sends two long streams of fuel out of the carburettor immediately and for a full one second duration. **Warning!** Take appropriate fire precautions when testing accelerator pump action and protect your eyes and skin too.

Chapter 7

Rolling road tuning & problem solving

The ultimate test of any engine is how well it goes in its particular application. Competition brings out the best in engines and the worst in faults. That is because of the conditions and stresses and strains that the machinery is being subjected to. For example, hundreds of cars are taken to the track each weekend and they are driven around lap after lap misfiring and backfiring. The drivers invariably are mystified at the antics of the engine: after all, they drove the car to the track and all was well. Bonnets are up and, usually, the carburettors are being scrutinized even though they are probably not to blame.

The real problem is usually one of preparation. Track conditions always show how well the work on the engine has been done. Sometimes, what should be really good engines have been run for years with what is, eventually, proved to be quite minor tuning problems.

For an engine to give top performance, every component must be proved to be in a serviceable condition. That goes for new parts too: never assume that because a part is new that it's good. Even top quality parts can be proved faulty by the stress of competition work but these same parts could be put into a standard road-going car and operate perfectly. The conditions in competition are just not the same. Problems often stem from spark plugs. Plugs used on a road-going car have no place on an engine being raced. Take a perfect set of plugs with you and fit them at the track but only when the engine is going to be subjected to high rpm (no idling and so on). Remove them from the engine when the racing is finished and pack them away until next time. If a spark plug is dropped to the ground don't never use it in a racing engine.

During testing and final adjustments, the question of choke sizes may arise, particularly whether there might be any power benefit derived from fitting bigger chokes. By all means increase (or decrease) the size of the chokes by 1mm at a time; however, it's unlikely that deviations of over 2mm from the recommendations given in chapter 4 will prove successful. Note, too, that if choke size is altered, the main jet/air corrector combination and idle jet will have to be rechecked in the original sequence and may need to be altered (richness/leanness factor). Chokes that are too large for the application will not allow the engine to accelerate as well as it would with smaller chokes, even though the maximum power output will be similar. Chokes that are too small will limit maximum rpm quite sharply. Taking the trouble to make sure that exactly the right sized chokes are fitted to any engine, that is the smallest that allow the engine to develop maximum power, pays off in no uncertain terms.

ROLLING ROAD (DYNO) TESTING PROCEDURE

Rolling road dynos are excellent pieces

of equipment that allow scientific diagnoses to be made, and power readings to be taken in simulated road/track conditions. Changes can be made and evaluated almost immediately, and the air/fuel ratio and ignition timing can be checked as you go. Because the testing is carried out in workshop conditions, all manner of useful tools will be at hand, and time will be saved over direct road or track testing. You can even look into the carburettor tracts as the engine is running. In the vast majority of cases, the car will prove to be 'right' on the road or track as soon as it comes off the rolling road.

Rolling road calibration does vary a bit, with some equipment giving higher readings than others. This is no cause for concern, provided you use the same dyno for all of your testing and setting up. Knowing the exact power output of an engine is, in the final analysis, not overly important, although it can, of course, tell you whether expectations on the required power output have been met. Most rolling roads are fairly accurate in the readings that they give.

A lot of people place a huge amount of importance on the maximum power output achieved, often to the exclusion of any other factor (mid-range power output and engine response throughout the rpm range, for example). Race winning cars generally have engines that perform well in all areas of the useable rpm band, and that means that the engine has been thoroughly tested throughout the working rpm range and adjusted as necessary.

Fan cooling is required when a rolling road is being used, and the 'driver' needs to keep a very watchful eye on the water temperature gauge as it's very easy for this aspect to get out of hand. Testing a car on a rolling road dyno makes it only too clear just how much air actually passes through the radiator to keep the engine temperature within the required limits when on the road or track.

Make sure that the vehicle is well secured. The non-driven wheels need to be firmly chocked, and the driven wheel end of the car must be tied to the floor by some means for lateral stability. Cars can and do get seriously out of control on rolling road dynos, and a lot can happen in an instant, with serious accidents possible. Take no risks here; it's not impossible to have a car effectively 'travelling' at 100-160mph/160-260kmh on a rolling road.

The first thing to do when testing a vehicle on a rolling road is to set the static ignition timing to about 10 degrees (that's 10 crankshaft degrees). This can be done fairly accurately with a points ignition, and by 'near enough' guesswork with an electronic pickup ignition. Most engines will start at this amount of ignition timing. If the engine has a vacuum advance mechanism, disconnect it, and temporarily plug the inlet manifold connection (with a rubber cap).

Start the engine and see if it will settle at a steady idle. Adjust the throttle so that the applicable rpm is achieved, and adjust the idle mixture adjustment screws so that as smooth an idle as possible is achieved.

Next, connect a strobe light to the engine and see what the ignition timing actually is. Set the ignition timing to exactly 10 degrees by this method, and then lower the rpm by 100-200rpm and see if the number of degrees of the ignition timing reduces. If it remains the same, then the mechanical advance has not started to operate at the recommended idle speed. If it reduces, then the mechanical advance mechanism was in operation and the advance springs in combination are too weak, and one or both will need to be changed now. Few engines need to have the mechanical advance mechanism start to work before 1500rpm is reached and, in fact, it's almost always better that it doesn't (however, it isn't always possible to have it like this). Slowly increase the engine speed and make a note of the rpm at which the ignition timing starts to increase (it might be important to take it into account later).

With it proved that the mechanical ignition advance is not operating before 1500rpm, the idle speed ignition timing can now be increased from 10 degrees to 16 degrees in 2 degree increments to find the optimum amount. Expect a standard engine to idle the smoothest at between 10-12 degrees of ignition timing, a 270-280 degree camshaft-equipped engine 10-14 degrees, and a racing engine at 12-16 degrees. The ideal amount ignition timing is that which causes the engine to idle the smoothest at the generally accepted idle speed. Advance the ignition timing from 10 degrees to 12 degrees and note whether or not the idle speed increased. If it did, slow the engine to the recommended idle speed for the type of engine and its state of tune, by adjusting the throttle adjustment screw. Next, adjust the ignition timing to 14 degrees and note whether or not the idle speed increases. Some engines' idle speeds won't increase by going from 10 to 12 degrees, but most will. Some engines won't increase speed going from 12 to 14 degrees, but some will. If the idle speed keeps increasing, slow it down to the recommended idle speed and advance the ignition timing to a maximum of 16 degrees.

Most engines will require an amount of idle speed ignition timing of between 10-16 degrees.

Some engine designs have a very narrow range within which they idle very smoothly, between 10-12 degrees, for example. Others have a wider range, such as 10-16 degrees. When the amount of ignition timing is too little the engine speed reduces, and when the ignition timing is too much the engine starts to idle roughly. The optimum amount of ignition timing is very often the mean amount between these two limits. If idle smoothness is very good at 14 degrees, for example, and idle speed is 1000rpm, increase the ignition timing to 16 degrees and see if the rpm increases. If it doesn't, then 14 degrees is going to be about right. Reduce the ignition timing to 12 degrees and see if the rpm reduces and the idle becomes less smooth. If the rpm reduces, 12 degrees is too little and 14 degrees is definitely the right amount on this basis. An occasional misfire, such as two or three times within a 10 second time frame, means that the ignition timing is either too advanced or not advanced enough (assuming the mixture strength is correct). It doesn't take too much effort to narrow down what the optimum static/idle speed ignition timing is.

Some engines like more than 16 degrees idle speed ignition timing. The Ford Pinto engine, for example, has always responded to 18-20 degrees of idle speed ignition timing. The problem that crops up when an engine needs this amount of static/idle speed ignition timing is that the starter motor very often won't easily turn over the engine without 'kicking back' (it's difficult to get the engine to turn over and start). Starter motors will usually turn engines over at a speed of about 500rpm, and virtually all starter motors will turn an engine over with 10 degrees of static ignition timing (no 'kicking back') and start it. In fact, most will turn an engine over with 14-16 degrees of static ignition timing. It's after about 16 degrees of static ignition timing that many engines will often start to 'kick back'.

It's quite easy to find out what the ignition timing limit is for starting a particular engine. The ignition timing is increased upwards from 10 degrees in 1 degree increments until there is resistance to starting. When the point of resistance to starting is reached, the maximum static ignition timing that can be used is 2 degrees less. It has to be 2 degrees less than the amount that causes the 'kicking back' otherwise it can become a bit marginal as to whether the engine will turn over at a reasonable speed and start. When the point of too much ignition timing has been reached, the distributor body is repositioned in the previous setting (retarded).

On engines that allow a maximum of 14 degrees of static ignition timing for starting purposes, for example, but need 20 degrees of idle speed ignition timing at 1300rpm, a mix of advance mechanism springs is going to be required. It takes a bit of work to achieve this but it is possible. With the static ignition timing set to 14 degrees, mark the distributor to block/cylinder head position for accurate replacement, remove the distributor, remove the strong advance spring and refit the distributor, carefully aligning the distributor marks so that the static timing is maintained. The engine is then started, but not revved. With the engine idling at the prescribed amount, the ignition timing is checked. If it's more than 20 degrees, the advance spring is too weak, and if the ignition timing is less than 20 degrees, the advance spring is too strong. Substitute the single spring until the ignition timing is as required. Some distributors have fixed spring posts while others have one post per spring that can be bent outwards or inwards to alter the tension. The latter are the easiest to adjust.

With this aspect of the static and idle ignition timing finalised, your

Blanking plate and gasket to cure fuel leakage.

engine will be starting at 14 degrees and idling with 20 degrees. With the engine idling at the prescribed idle speed and the ignition timing set as required, connect the strobe light and slow the engine to see if the static amount of ignition timing is being realised (this applies mainly to electronic ignitions which can't be set like points). The object of the test is to see if, when the mechanical advance mechanism is back on its stops, the engine is actually at the ideal number of starting degrees. The engine may stop when you do this.

The next stage is to fit a second spring which allows the ignition timing to be advanced at a suitable rate from the 20 degrees to maximum mechanical advance, so that the maximum amount of ignition timing is 'all in' just on or before maximum torque is reached. Maximum torque will only be realised if the optimum amount of ignition timing is present. Such springs have to be fitted so that they exert no tension until the mechanism moves beyond the 20 degree point. If this doesn't happen, only the first spring will be controlling the rate of ignition advance and, being quite a weak spring, it will allow too much ignition advance too quickly (which can lead to reduced acceleration compared to what is possible).

Many distributors feature one spring which is loose fitting when the advance mechanism is in the static position. As the mechanism advances these springs come into play. Standard fitted springs like this are usually quite strong springs but they don't have to be. Mixing springs is a way of gaining a non-linear advance rate. The second spring must allow the mechanical advance to increase at the ideal rate. Too quick, and the engine will tend to 'kick back' and an audible knock will

probably be heard under acceleration. Increase the second spring's tension immediately. There is a lot you can do with the advance springs to obtain 'the perfect' ignition timing regime.

Engines equipped with conventional distributors do not accelerate using ignition timing 'put into' the engine via the vacuum canister/vacuum advance mechanism. The vacuum advance only works when vacuum is being generated by the engine, such as at cruise (partial throttle). In such instances the vacuum advance will 'put in' another 10-12 degrees of total ignition timing. The instant the engine goes off vacuum (accelerator pedal depressed to wide open throttle) the vacuum advance canister 'pulls' the advance plate back and the engine accelerates using the mechanical advance only. If the vacuum advance always seems to be 'working', it is doing so because the particular vacuum canister is calibrated to receive partial vacuum from the carburettor as opposed to inlet manifold vacuum. It's virtually impossible to reduce inlet manifold vacuum to suit a vacuum canister designed for partial vacuum. Overall, the ignition timing curve is set without the vacuum advance mechanism in operation, but you do have to check that the vacuum advance is not in operation at idle too much due to an incorrect connection take off point.

Air/fuel mixture ratios
A lot of people have trouble getting the air/fuel mixture ratios right, especially at the 'top end'. Having an engine 'run up' on a rolling road with Lambda or percentage CO exhaust gas analysing equipment connected can solve all of this and give you piece of mind that the mixture is not going to be either too lean or too rich. In fact, the top end mixture strength can be

set very accurately on a rolling road. What happens is that the engine is run at wide open throttle under load on the rollers, a maximum power reading is taken and the exhaust gases analysed. The mixture strength is then either increased or increased, and the engine once again run at wide open throttle and the power reading and mixture strength recorded. The two results are then compared. If the power increased at the second test, the engine needed increased mixture strength. If the power reduced at the second test the mixture was too rich. The mixture strength is then reduced to find out when the power starts to reduce. It only takes a few runs to work out what the mixture strength needs to be to produce maximum power. For example, if an engine produces maximum power of 200bhp with a air/fuel mixture ratio of 0.85 Lambda, 12.5:1 air/fuel or 5.0% CO, it might produce about 195bhp with a 0.84 - 0.86 Lambda or 5.3 - 4.7% CO reading, which is a plus and minus reading about a mean optimum figure.

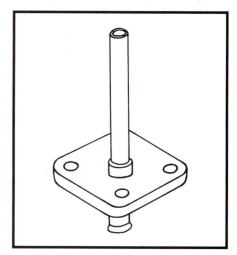

Tube fitted into a mounting plate. Tube is 0.312in/8.0mm in diameter and has had the top 0.62in/15mm turned down to 0.30in/7.5mm in diameter. The tube protrudes 2.51in/64mm from the underside of the mounting plate.

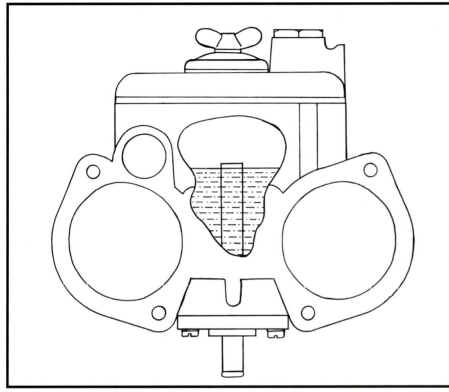

Above - fuel drain installation. Below - Schematic of full scavenging system showing how excess fuel is pumped back to the fuel tank.

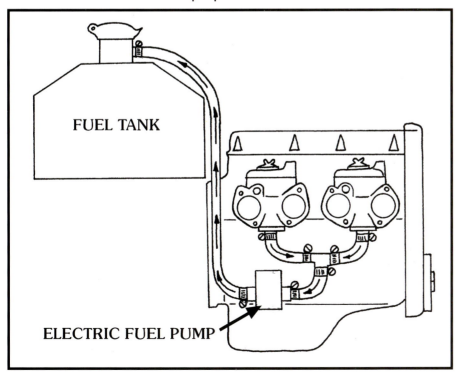

FUEL TANK

ELECTRIC FUEL PUMP

This is how you obtain the optimum air/fuel mixture for maximum power production.

The same applies at maximum torque. The mixture is set plus and minus of the optimum amount to find the optimum. You don't know what the optimum fueling for maximum torque is until the power 'falls off' through making the mixture too rich. Once the foot/pounds reduces after you have made the mixture too rich you go back to the previous fueling amount. The optimum fueling is that which generates the maximum foot pounds of torque figure. It doesn't actually matter what it is. Ignition timing and fuel mixture requirements of an engine are very closely related in this sense It's actually a very straightforward procedure to find an engine's ignition timing and fueling requirements.

Caution! The recommendations given in this book are of a general nature. It assumes that your engine is not over compressed for the octane rating of the fuel being used. If you are using an octane rating of fuel that will not support the compression ratio you are using, engine damage is possible (burnt pistons, for example.). It's quite possible to achieve maximum power on such engines, but not reliably because the internal componentry runs too hot. In some cases a slightly richer mixture than the optimum required to produce maximum power is necessary to keep the internal componentry cool. The ignition timing might also have to be 'backed off' (reduced). Because it's not always just a case of setting an engine for maximum power and choosing the correct mixture, the recommendations listed in this book are theoretical.

Note that the amount ignition timing plays a huge part in achieving optimum engine power and efficiency.

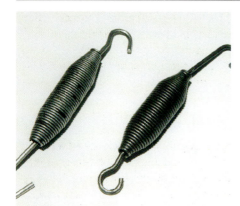

Return springs made specifically for use with Weber or Dellorto carburettors.

Insufficient ignition timing is known to cause overheating, and too much ignition timing causes a power reduction and poor acceleration. Many perceived problems with engine and carburettor tuning problems especially can often ultimately be traced to the ignition system!

As a general rule the ignition timing is set at the point of maximum torque, this being the point of maximum charge density. On wide open throttle with the engine under load at maximum torque, the ignition timing is advanced one degree at a time to see how it responds. When the amount of torque stops increasing the ignition timing is stopped being increased. You do have to 'over advance' the ignition timing by 1-2 degrees to find the ideal amount of ignition timing, of course. The number of degrees of ignition timing and the air/fuel ratio in combination play their part in achieving maximum torque. Get one wrong and the engine will be down on efficiency. If an engine 'pinks', check the air/fuel ratio, just in case it's lean, and if there is a fuel supply problem. Too much ignition timing and the engine will 'pink', too little ignition timing and the foot pounds of torque will reduce. Careful adjustments and checks have to be made to make sure that the air/fuel ratio is correct at all times and the ignition timing adjusted to suit.

After the point of maximum torque has been reached, the amount of ignition timing is generally academic because the volumetric efficiency starts to reduce and the prospect of 'pinking' reduces along with it. The maximum amount of ignition timing required is that which causes maximum torque to be produced. With this found the amount of mechanical advance the distributor requires is then built into it. This is done by changing the length of the mechanical advance slot in the base plate or finding a factory advance plate which has the correct amount of mechanical advance allowance. If the advance plate has too long a slot in it, this can be welded and re-shaped to make it shorter. If an advance plate with the correct sized slot is not available, you'll have to weld a standard advance plate. Mallory (USA) makes distributors which come with an adjustable advance plate. It takes 5 minutes to alter the length of the advance slot to the precise required length. These distributors are of excellent quality and ideal because of the adjustment factor. Note that if an advance plate is going to be welded, you would normally weld it at the full advance end of the slot. If you weld at the 'at rest' position end of the slot, all of the advance spring settings will be altered because the advance plate will no longer be in the original position.

Most standard road going engines produce maximum torque between 2400-3800rpm, or at about the halfway point in the rpm range. For example, if an engine manufacturer lists the maximum power as occurring at 7000rpm, it's highly likely that maximum torque is going to be produced around 3500rpm, or between 3000-4000rpm. A lot of modern 4 cylinder, in-line engines are like this. In days gone by, American V8 engines might have developed maximum power at 4400rpm, and often maximum torque was developed at 2200-2400rpm. Be aware of the fact that some engines 'like' a few degrees (2-4) more or less ignition timing than that required to produce maximum torque, at or near maximum rpm, to produce best possible top end power. The increase in power is not much but it's usually worth having.

With programmable electronic engine management systems it's quite possible to increase or decrease the amount of ignition timing anywhere in the rpm range, and 'pick up' this 'lost power' if an engine does have such a requirement. This is less easy to build into a distributor. Today some aftermarket electronic ignition systems that can be used with carburettors have the facility to make such adjustments.

Remember to have the electrics of your engine checked for integrity. The coil (even a brand new one) should be tested for output and uniformity of spark production. The spark plug wires should be checked for resistance. Make sure that the amount of spark being delivered to the spark plug is as prescribed by the manufacturer. Another book in Veloce's SpeedPro Series (*How to build & power tune Distributor Type Ignitions Systems*) covers the sort of checks that have to be made and gives guides to all of the requirements.

TRACK TESTING PROCEDURE

The ultimate test of any tune-up is how well the engine runs on the road or track. But what goes well on the road may not go well on the track. This is because on the race track the engine will be operating from mid-range up to maximum rpm. All sorts of things happen in a track environment that will seldom happen elsewhere. For example, insufficient fuel pressure or volume will cause the engine to miss as the fuel level in the carburettors falls. The high load and sustained high rpm of track life will show up problems that would not be encountered on the road. Another example is that pulling out of turns using full throttle can make an engine hesitate and misfire, yet before the vehicle was tried on the track it was perfect. What all this means is that carburettor set-up may have to be altered to compensate for the demands of full bore track work. Luckily Dellorto and Weber carburettors are totally adjustable for all conditions and applications. However, they can only be made to suit one application at a time!

When track testing, use top gear to check the top end performance and this is achieved by running the car over a set distance using top gear only. If the engine has a maximum rpm of 7000 and is on the power band (camshaft working) at, say, 3300rpm check acceleration over two set points. When the first set point (flag on a stick at the side of the track) is reached the engine must be at 3500rpm and in top gear. At the set point depress the accelerator to the floor and leave it there until the second set point (another flag on a stick at the side of the track) is reached. At the second set point note the rpm. If at the first trial the rpm was 6600rpm this is the datum against which all future adjustments are measured. The second set point can be moved (if space allows) so that the engine just makes maximum rpm. Also from the first set point a stop watch should be used to check the time taken to get from the first set point to the second set point. Note the rpm and the times. Make only one engine adjustment or change at a time and

measure the value of each change by the rpm and the time taken over the test distance.

When testing such as this is carried out, the engine is under the highest load possible and the rpm range is more than would usually be used. If the engine is going to falter it will do so under these conditions. There are no gearchanges to make (they could interfere with the consistency of the results) and the flexibility of the engine is fully tested.

SOLVING PROBLEMS - LOW TO MID-RANGE RPM

If the main jet is too small the engine will not produce good power anywhere in the range. An excessively lean overall mixture will see the engine missing and backfiring through the carburettor chokes under load. Continue to increase the main jet size as long as the power keeps going up.

For 'flat spots' or hesitation check the accelerator pump jet sizes. Firstly, increase the pump jet by up to 5 and see if this improves the situation. If this works to a very small degree and the misfire or hesitation is still present but now slightly higher up in the rpm range, the problem is not with the accelerator pump jet, so go back to the original size. What is happening in this scenario is that the accelerator pump is masking the real problem. Next, change the emulsion tube to a richer one (see chapter 4 for details).

If the emulsion tube is wrong for the engine it will show up during acceleration. With the engine in top gear, accelerate the engine slowly from 3500rpm (or from the rev range at which your engine starts to make power, comes 'on cam'), slowly so that the engine does not get a full shot of fuel from the accelerator pumps. If the emulsion is wrong the engine will 'stumble' then clear as the revs rise.

This 'stumble' is caused by acceleration phase leanness and a richer mixture can be provided by changing to an emulsion tube that will supply a richer air/fuel mixture.

SOLVING PROBLEMS - HIGH RPM

If the engine has a high speed misfire reduce the size of the air corrector by 20, 40 and 60 but not to less than the size of the main jet (if the main jet is a 160, don't go less than 160 for the air corrector). If this fails go back to the largest air corrector that you used previously and increase the main jet size by 5 and then reduce the sizes of the air correctors again.

Check the total spark advance again and increase it by 2 degrees for the purposes of the test. If further spark advancing is contemplated do not persist with high rpm and over advanced ignition timing (no more than 5 seconds under full throttle when the engine is misfiring). If the engine responds to the increased spark advance, go back and check the TDC position marks and the advance degree markings for accuracy before continuing running the engine with what could be excessive total spark advance (engine damage could result).

Hook up a rig and check the fuel pressure at the time of engine misfire. Check all ignition components (even if the parts are new). This is often best done in conjunction with an auto-electrician who will usually have the latest test equipment. Avoid prolonged running of the engine up to full rpm with no load on it (free revving). This serves no useful purpose but go high momentarily to see if any breakdown of the spark can be detected. If an engine has a misfire in the free revving situation (no load) it will most certainly have the miss - and worse - under load.

Failing this the vehicle will have to be put on a rolling road (which puts the engine under load) and the engine run up with the electrical testing equipment connected to it. This will remove all doubt as to the integrity of the electrical system. Things like plug leads and spark plugs (even if new) can cause problems and this is the quickest way to isolate and fix an electrical problem.

WEBER - FUEL LEAKAGE FROM FUEL ENRICHMENT DEVICE

When sidedraught Webers are inclined to any degree there is always the possibility that fuel will leak out of the fuel enrichment device. Even if the starter valve holes into the actual chokes are blocked off to prevent internal fuel leakage, there is still the prospect of external fuel leakage.

This problem usually applies to older, well used carburettors. Warning! - While such fuel leakage is not all that common, it does occur and, especially if the exhaust pipes are on the same side as the carburettors, something must be done to prevent fuel leakage. If fuel is found to be leaking out of the fuel enrichment device the following is the solution to this problem.

With this particular problem there is no possibility of a repair and so, as the fuel enrichment device mechanism is not necessary, it can be removed. To stop the fuel leakage from the back of the carburettor the fuel enrichment mechanism is removed and a new flat plate made up the same size as the original and secured on to the back of the carburettor as per normal.

The replacement plate is made out of 5mm/0.187in thick aluminium flat sheet to the exact outline and size as the original enrichment device. The holes for the securing screws are drilled in the same place. A gasket is made up and fitted between the plate and body of the carburettor. The gasket is made out of 0.020in/0.50mm thick gasket paper.

When the plate and gasket are securely in place there will be no more leaks.

WEBER - ADAPTING FOR OFF-ROAD APPLICATIONS

Many engines are built for off-road use where the terrain is very rough. The result of using either Webers or Dellortos on engines in such situations results in one thing - FLOODING. The definition of flooding being, of course, that the carburettor temporarily ends up with a much higher fuel level than it is designed to run with, resulting in a very rich mixture.

All manner of things have been tried over the years in an effort to cure this problem but, until now, none have ever worked satisfactorily, to the point that everyone who uses these carburettors over rough terrain just lives with the problem.

This was not acceptable to one of my customers about 20 years ago and he asked me to see what I could do about the problem as he would have an advantage without the flooding. It took ages to work out but it was worthwhile as at the time he was virtually unbeatable in his class.

This situation has gone on now for so long that if someone told you that you could get Webers to run like a fuel injection system, you'd be hard pressed to believe them. The fact is, however, that Webers can be easily altered to run over any terrain without any problems. It all seems too good to be true but, nevertheless, the suggested modifications are realistic, practical and work exceptionally well.

Webers only are described here because their construction allows the easy fitment of the custom made accessory needed for the transformation. The Dellorto is constructed slightly differently and is, as a consequence, not as easy to alter - but can be altered by more drastic means to achieve the same result.

What happens with these carburettors, when they are used in off road applications, is that the floats/needle and seat allow too much fuel in. When using the standard float system there is no way that this can be prevented. If too much fuel gets into the fuel bowl, the engine will have a rich mixture and will only run correctly again once it has used the fuel in the float chamber to the point that the float level becomes correct again. In the meantime, the engine will be flooded, will 'miss' and will not produce full power.

The original floats, needles and seats meter the fuel into the float chamber perfectly, of that there is no question. The problem is that because of the angles that the carburettor/s are placed at over rough terrain, the float heights don't truly represent the amount of fuel in the float chamber and the needle valve lets more fuel into the chamber. Reducing the sizes of the needle and seat will reduce the amount of fuel ultimately allowed into the float chamber, but will not prevent excess fuel being allowed in. The fact that the float and needle valve is 'fooled' by the gross angle the carburettor is inclined at a particular moment is a problem about which nothing that can be done, but there is something that can be done about removing excess fuel.

The solution to this problem of excess fuel in the float chamber is to remove the excess fuel from the float chamber virtually instantly. This is achieved by rigging up a centrally situated tube in the float chamber to remove excess fuel. An 0.236in/6mm, or better still 0.312in/8mm, diameter metal tube is fitted in the main jet tower of the carburettor. The top 0.62in/15mm of the tube needs to be turned down to 0.30in/7.5mm to clear internal casting shape. The top of the tube will be level with the surface level of the fuel in the float chamber when the carburettor is on level ground. This means that the top of the tube must be 2.51in/64mm above the mounting plate. The inside diameter of the tube needs to be a minimum of 0.187in/5mm to a maximum of 0.236in/6mm for tube sizes suggested.

The original bottom well cover is replaced with a similar shaped piece of 0.312in/8mm thick aluminium which has the thin walled tube fitted into it. The end result is a drain positioned in the centre of the carburettor's fuel bowl. When the fuel level rises too high, the excess is immediately sucked away and returned to the fuel tank.

If the fuel tank is lower than the carburettors the fuel could self-drain out of the carburettor float chambers and back into the fuel tank, but off-road machines almost always have the fuel tank higher than the carburettors. Therefore, to scavenge the excess

fuel, a small capacity Facet electric fuel pump, or similar, is used. The fuel pump can be fixed to run when the ignition is on, or set up to be switched on when the vehicle encounters rough terrain. Whatever the height of the fuel tank, the use of a fuel pump is the best option because excess fuel will be removed faster.

Theoretically, it's possible for even a small pump to suck air from the float chamber to the extent that there is no longer sufficient atmospheric air pressure acting on the top of the fuel. The carburettor is then operating with less than 14.7 pounds of air pressure acting on top of the fuel. If this happens, the engine will not run properly. However, the float chamber air vent of most Webers is very large and it's unlikely that any pump would be able to reduce air pressure within the float chambers to below ambient.

It will, of course, be necessary to connect the overflow tube to the fuel tank, so that scavenged fuel is returned safely to the tank and the whole system remains sealed.

The carburettors can quickly be returned to stock condition if required.

RETURN SPRINGS

Both Dellortos and Webers have one return spring per carburettor, but it is never acceptable to rely solely on the carburettor's own return springs. No race organisation will allow an engine to be used in an organised event with just these return spring/s alone, and quite rightly so. Regardless of application, a minimum of two extra return springs must be used on any engine.

MAINTENANCE

Once the carburettors have been set-up to give optimum performance, you'll still need to check them occasionally to keep them working at their best. This checking will involve throttle spindle synchronization, float level checking and possible re-setting (note that rough terrain work will upset the float levels very quickly) and idle mixture screw adjustment, ensuring the throttles open fully and changing the fuel filter at regular intervals.

Chapter 8

Fuel & octane ratings

The vast majority of fuels commercially available around the developed world these days are unleaded, and the octane rating of these fuels ranges from 90 to 98.3 RON (as of 2005). The last of the tetraethyl-leaded fuels were, on average, 93-98 RON octane, but they have now gone from all developed countries of the world, even for motor racing purposes. There are two basic tests used worldwide to determine the octane ratings of petrol/gasoline used in cars. These two test methods are the Research Octane Number (RON) and the Motor Octane Number (MON).

In the late 1920s, the Co-operative Fuels Research Council (CFR) in the USA devised a fuel sample testing method for determining the octane ratings of petrol/gasoline mixes. A very special single cylinder engine was made which had variable compression ratio and ignition timing (commonly referred to as a CFR engine). This very same engine is still used to test fuel samples for RON and MON using two different testing regimes or methods. Before a fuel sample is tested, the CFR engine is calibrated using a pure chemical mix 'reference fuel', which, because of its specific chemical content, is guaranteed to be 100 octane.

A sample of fuel to be tested is then used to run the engine. The RON method of obtaining an octane rating sees the CFR engine run at 600rpm with a set amount of ignition timing, as prescribed by ASTM D2699/IP237 (13 degrees Before Top Dead Centre). This rating is regarded as being representative of how the particular fuel will cause an engine to go at start up and idle. ASTM stands for American Society for Testing and Materials, and D2699 is the criteria of its RON test. IP stands for Institute of Petroleum, and 237 is its number for this testing regime (same test).

The same sample of fuel is again used to run the engine and the MON method of obtaining an octane rating is conducted under ASTM D2700/IP236 criteria. The CFR engine is run at 900rpm, the compression ratio is increased, and the ignition timing is advanced. The octane derived from this test is regarded as being representative of how the particular fuel sample will cause an engine to go on the road at cruise conditions or motorway driving. You virtually never see a MON rating displayed on a pump in a service station, but the two ratings are used by all petrol companies around the world. BS4040, for example, lists a minimum requirement for both RON and MON, but you only ever see the RON rating displayed on a pump.

At the time of writing (2005), the UK still has some leaded fuel available to motorists/enthusiasts. When tetraethyl-lead fuel was phased out in 2000, the British Government allowed 1% of the total amount of petrol made and sold in the UK to be leaded. The petrol companies which decided to

Exhaust valve seat in a head used with unleaded fuel for many miles: massive deterioration is obvious. All the exhaust valve seats in this head were the same. This is not valve seat recession, it is valve seat deterioration. Valve seat recession is the next stage, as the original valve seat has not been completely eroded away yet.

make this fuel and appoint selected outlets to sell it, used 0.099 grams per litre of tetraethyl-lead which provides acceptable valve seat protection and the fuel is 98 RON/86.2 MON.

Two unleaded fuels have been available in the UK for some time now, and these are Premium unleaded, which is required under BS7070 (British Standard 7070) to have a minimum RON of 95, and Super unleaded, which is required to have a minimum RON of 97. There is a slight problem with Super unleaded, however, if the petrol station holding it in its tanks is not a busy one. Super unleaded gets its extra octane by having various 'volatiles' mixed into it, and they tend to evaporate quite quickly. The longer the fuel remains in the tanks of the petrol station, the less the octane rating of the fuel. Premium unleaded does not have these 'volatiles' in it and, as a consequence, maintains its manufactured octane (95) rating longer. Because of this, Premium unleaded will, in some instances, cause your car to go better than

Super unleaded. As with all petrol/gasoline, if maintaining the octane is important (if you're using a high compression engine, for example), the fresher the fuel the better, and the recommendation is to buy your fuel from a busy forecourt. Old Super unleaded fuel can end up with a lower octane than Premium 95 in certain circumstances. The highest octane unleaded fuel commercially available in the UK today (2005) is Shell Optimax, which is 98.3 RON and 86.9 MON, BOP also has a similar fuel available.

The USA, while still using the RON and MON tests to rate fuel, has taken this all one step further by introducing an Anti-Knock Index (AKI) number, based on the RON and the MON added together and then divided by 2. Other names for this system you see used in the USA are PP (Pump Posted) or perhaps PON (Pump Octane Number). As you can see, when you're talking octane, you need to be quite clear what criteria are being used.

In the USA there are three basic

The valve on the right came out of the combustion chamber shown on the previous page, and the valve on the left came out of an adjacent combustion chamber with a similar seat condition.

grades of AKI street legal unleaded fuels on sale at the pumps, with slight variations in the Premium unleaded fuels, plus a very low octane Regular unleaded (New Mexico 85 AKI octane). For the purposes of matching the fuel available to a compression ratio that will be suitable, use the RON ratings of the US fuels listed. Take it that, on average, Regular unleaded 87 octane is suitable for use with 8.3-8.8:1 compression engines, Mid-grade unleaded 89 octane is suitable for use with 8.8-9.2:1 compression engines, and Premium unleaded 91 AKI octane is suitable for use with 9.0-9.3:1 compression engines. Add 0.3 for 92 AKI octane and 0.7 for 93 AKI octane.

Regular unleaded   87 octane is 91 RON - 83 MON
Mid-grade unleaded 89 octane is 94 RON - 84 MON
Premium unleaded   91 octane is 96 RON - 86 MON
or   92 octane is 97 RON - 87 MON
or   93 octane is 98 RON - 88 MON

For high octane street legal unleaded fuels, and for 'off road' racing purposes, USA enthusiasts are fortunate compared to those in other countries around the world, because there is a vast array of specially mixed and blended fuels available (for details of VP products go to www.vpracingfuels.com; the products of Sunoco can be viewed at www.sunoco.com; for F&L Racing Fuel products go to www.fandl.com; and for Sports Racing Gasoline products see www.cosbyoil.com.

If you look at the USA's Sunoco GT100 street legal unleaded fuel, for example, which is advertised as being 100 octane, is 105 RON and a 95 MON, making it a pretty good fuel for road use. The 100 octane rating you see advertised is the AKI, which is 105 + 95 divided by 2 = 100. This fuel will run a naturally aspirated A-Series engine with a compression ratio of 12.5:1 without any problem. Sunoco also makes an unleaded fuel called GT Plus, which has RON and MON values of 109 and 99, respectively, giving a 104 AKI, and will run a racing

engine with a compression ratio of 14:1 without problems. VP C10 Performance Unleaded is an equivalent fuel to Sunoco GT100, and the street legal VP Motorsport 103 is equivalent to Sunoco GT Plus. USA owners of 'soft' A-Series cylinder heads would need to use a lead substitute additive with these fuels to prevent valve seat recession.

Sunoco makes 'Standard Leaded' for racing purposes, and this has RON and MON ratings of 115 and 105 MON, respectively, and 0.99 grams of tetraethyl-lead per litre. Next is 'Supreme Leaded', which has RON and MON values of 114 and 110 MON, respectively, and 1.12 grams of tetraethyl-lead per litre. Finally there is 'Maximal Leaded' which has RON and MON ratings of 118 and 115, rexpectively, and 1.32 grams of tetraethyl-lead per litre. VP and Sports Racing Gasoline make equivalents. All of these specialist racing fuels are excellent, but they are quite expensive.

For the benefit of older cars not designed to run on un-leaded fuel, Britain introduced lead replacement petrol (LRP) in 2000, when tetraethyl-leaded fuels were phased out. LRP was 95 RON octane unleaded fuel with potassium added at the rate of 8 parts per million. It was claimed that LRP ffered near equal valve seat recession protection to that of tetraethyl-lead fuel. Certainly, for all round general use, engines did not suffer from valve seat recession while using it. LRP was not overly suitable for a high load situations, however, such as when an engine is driven at wide open throttle for long distances, though the petrol companies did state this at the time of the fuel's introduction. LRP fuel is not as common today as it used to be, and likely to be phased out completely within a few years.

HEADS WITHOUT HARDENED EXHAUST VALVE SEATS

Some countries still offer leaded fuel. The UK, for instance, still allows the production and sale of leaded fuel to garages that are prepared to stock it and sell it, on the basis that the amount of fuel sold nationwide is not more than 0.5% of the total amount of fuel sold. This means that leaded fuel is reasonably readily available for those people prepared to pay a premium price for it.

At 0.149 grams per litre, the amount of tetraethyl-lead in this fuel is quite small, but it did comply with BS4040. In the days of 5 Star fuel (mid-1950s, 1960s and 1970s) there was up to 3 grams of tetraethyl-lead per gallon. By the time 4 Star fuel was phased out, however, in 2000, BS4040 called for the amount of tetraethyl-lead content per litre to range from 0.05 to 0.150 grams. The major fuel companies opted for 0.149 grams per litre, keeping the tetraethyl-lead content right on the limit, in an effort to produce the best fuel possible within the confines of the standard. Many people may recall that some UK supermarkets were selling cheap fuel in the 1990s, with 0.05 grams of tetraethyl-lead per litre and approximate 92-93 RON octane rating. Anyone who used this fuel may have noticed getting a bit more fuel for their money, but probably a 20-25% drop in fuel economy. This was the very minimum of tetraethyl-lead allowed within the BS4040 standard, but was just enough by all accounts to stop valve seat recession.

Over a period of 5 years, I have conducted a series of experiments

This cylinder head has about 1/16in/1.5mm of valve seat recession. All you can do now is fit a hardened valve seat insert. It's quite 'fixable'.

using unleaded fuel in two BMC/Leyland/Rover A-Series all cast iron cylinder head engines which didn't have hardened valve seats fitted. In each case, the valve seats had done a lot of service, but the valves were still in good condition. Although the valve seats were regularly inspected, they were not reground or refurbished in any way. With both cylinder heads already having covered about 75,000 miles/120,000km on leaded fuel, they were put back into service. Both engines ran 10,000 and 25,000 miles/16,000 and 40,000km, respectively, and neither cylinder head failed in the more-or-less identical light duty applications (2500-3500rpm non-motorway applications). The 1275cc cylinder head, however, did show major signs of deterioration after being subjected to 10,000 miles/16,000km. The 998cc engine was subjected to the very same sort of treatment (same driver doing the same thing day after day) and did 25,000 miles/40,000km.

The valve seats in this engine's cylinder head were virtually the same as they were at the start of the test and could have gone on. There was some deterioration, but it was minimal, and followed the usual pattern for these engines.

Both engines had efficient cooling systems and were operating at similar water temperatures (78-80 degrees C/172-175 degrees F). This is an important point, as high engine operating temperatures make a difference, as the higher the temperature the worse the deterioration). The 998cc cylinder head, with more material around the exhaust valve and an engine which generates less heat, fared much better than the 1275 head.

Caution! - An engine which doesn't have hardened exhaust valve seats fitted to it must be run on either lead-replacement fuel (LRP), tetraethyl-leaded 4 Star fuel if it's available, or unleaded fuel with an additive mixed with it. There is some variation in the effectiveness of the approved additives which are available, but none are as good as having the known recommended amount of tetraethyl-lead in the fuel. The approved additives offer a measure of protection and, for general road going driving, they all seem to offer sufficient protection. Some have octane boosting properties as well.

Caution! - The use of unleaded fuel in engines without hardened valve seats will result in deterioration of the seats and valves at a rate commensurate with the way the car is driven. Even a car driven around town all of the time at slow speeds and low engine loading will show some deterioration over, say, 10,000miles/16,000km, but the damage will not necessarily be major and the valve sealing not necessarily impaired. However, take the same car and travel at full speed for hour upon hour and the valves will quickly start to recess. Expect major damage after 2000miles/3200km of this sort of treatment. Expect the deterioration to start immediately with this sort of treatment but not necessarily show up until 2000miles/3200km have been covered (perhaps a bit more). Expect an engine which already has poor exhaust valve seating to fail much quicker. It's the heat that does the damage ...

The range of valve seat recession prevention additives suitable for road going applications and mild competition use includes Superblend Zero Lead 2000, Redex 4 Star, Wynns, Millers VSP Plus, Valve Master and Red Line. There are plenty of others but none of them seem to be capable of being used in a real heavy duty application like motor racing. The only exception is TetraBOOST, which is not in quite the same category as the other additives. Anyone using an existing modified cast iron cylinder head, which has big valves, and which can't have hardened exhaust valve seats fitted to it, needs to add TetraBOOST to the unleaded fuel for maximum valve seat protection and/or its RON octane boosting properties. TetraBOOST has 1.23% tetraethyl-lead in it per litre, as well as 0.356% dibromoethane, and 0.385% dichoroethane, all in a solution of naphtha light aromatic. It's FBHVC (Federation of British Historic Car Clubs) tested and approved (2001). Mixed with 95 RON octane unleaded at the minimum recommended proportions, the combination becomes equivalent to 97 RON octane leaded fuel as per BS4040. Mixed with Shell Optimax 98.3 RON octane unleaded fuel at the minimum recommended proportions, the combination is equivalent to 100 RON octane leaded 4 Star fuel. Increase the amount of additive to 98.3 RON octane Shell Optimax, and the RON octane number increases to over 100, with up to 105-106 RON octane fuel being possible, but a lot of additive will be required to achieve this. You can get more information on TetraBOOST from Nik Cookson, TetraBOOST Ltd., 17 West Hill, London, SW19 1RB (Tel: +44 (0) 208 870 9933, Web: www.tetraboost. com.

The overall impact of the loss of leaded fuel is that regrinding valves and perhaps valve seats may end up becoming a more frequent practice, as might the replacement of exhaust valves. Engines with perfect valve seating produce optimum engine power and engines with anything less don't. Using a cylinder head which has pit marks in the exhaust valve seats is just no good at all as it leads to premature valve seat failure. Nothing can be 'got away with' when it comes to exhaust valve seating on engines: once a valve seat is damaged it usually takes quite a lot of regrinding to restore it. This can only be taken so far before either valve seat inserts have to be fitted, a slightly larger diameter exhaust valve has to be fitted, or the cylinder head is scrapped.

Fitting hardened valve seats
Fitting hardened valve seats is an option once the original seats are damaged and is, of course, the solution to being able to use unleaded fuel without additives. Unleaded cylinder heads are available as outright purchases from specialists or your own engine's cylinder head can have hardened valve seats fitted to it by your local engine reconditioner/machine-shop. Once a cylinder head has been fitted with hardened valve seats, unleaded fuel can be used without risk of valve seat recession. The fitting of hardened exhaust valve seats is a one off cost and being able to use straight unleaded fuel out of the pump makes for uncomplicated motoring. One thing that should always be done when converting your own cylinder head is to fit brand new exhaust valves.

While fitting hardened exhaust valve seats removes one problem, you do need to be aware of the fact that unleaded fuel is extremely hard on exhaust valves and no exhaust valve is going to last as long in an unleaded fuel engine as in a leaded fuel engine of years gone by. Expect exhaust valve contact area (seat) deterioration at a much greater rate than normal with 45,000 - 60,000miles/75,000 - 95,000km being the useful life of an exhaust valve. While it's possible to regrind the original valves, it's more prudent to replace them.

Engine reconitioners fit hardened exhaust valve seats to all non-induction hardened cast iron cylinder by boring the valve seat out so that there is a recess in the cylinder head. This recess is sized precisely to take a ready made hardened valve seat insert. The hardened valve seat insert is, on average, 0.004-0.006in/0.1-0.15mm larger than the aperture it is to fit into (an interference fit). The difference in size is dependent on the diameter of the insert. Engine reconditioners have a specification sheet that they use to make sure that they use the right size. After boring the recesses the diameter is then checked and so are the hardened valve seat inserts to make sure that the right amount of interference is available. The inserts are then carefully pressed into the cylinder heads. Alternatively, the cylinder head can be warmed up and the inserts frozen (freezer or liquid nitrogen), which allows them to be dropped into place.

Aluminium cylinder heads are always warmed up (150C/330F), and the frozen inserts carefully guided into place using a mandrel and a guide block. This way there is no possibility of the insert NOT going straight into the recess (can't 'pick up' on the side of the bored recess).

In all cases, the correct amount of interefence fit MUST be used.

CYLINDER HEADS WITH HARDENED EXHAUST VALVE SEATS

Caution! - If an engine fitted with hardened exhaust valve seats is excessively overheated (radiator failure for example) or a valve seat not installed in the cylinder head correctly, the seats can come loose causing engine damage. While this is unlikely, it is certainly possible ...

STICKING VALVES

With the widespread use of unleaded fuel throughout the developed world, many engines are now 'experiencing' sticking valves. With the exhaust valve/valves not shutting off properly, the engine loses power and starts to vibrate excessively. It can become so bad that a once powerful car is no longer able to climb a slight incline or maintain any sort of straight line speed. The vibration can become so bad that you might be inclined to think that the wheels are about to fall off. It's a quite extraordinary feeling driving a car with this problem.

The solution to the problem is to K-Line the valve guides or 'Tufftride' the valves. Either way, the cylinder heads have to come off. Straight unleaded fuel can be used after this action has been taken.

Chapter 9

Jetting/setting examples

The following starting point jetting/settings are for common engines/engine sizes. Ultimately, the optimal settings for individual applications will vary, but the range of jettings/settings listed, commensurate with the engine size and the degree of engine modification, will give a reasonable guide to what will ultimately be required.

Checking all engines using a gas analysing equipment is recommended, with with 0.85 Lambda/5.0% CO meaning that the engine has a 12.5:1 air/fuel mixture ratio, being the vital figure for full power production.

Note that Eurocarb Ltd (formerly Contact Developments) in the UK have a huge listing of sidedraft Dellorto jettings/settings for all manner of engines, in all states of tune. This company will supply spare parts and jetting information to anyone, anywhere in the world. They also supply Weber parts as well as Dellorto parts. Their address is -

Eurocarb Ltd,
256 Kentwood Hill,
Tilehurst,
Reading,
RG31 6DR
England,
Telephone 0118-943-1180,
Fax 0118-943-1190,
www.dellorto.com,
e-mail sales@dellorto.co.uk

BMC/Rover A-Series 1275cc engine (Weber)

The following settings were those arrived at by test on a 1293cc BMC A-Series engine. This engine uses readily available high performance parts, plus many standard parts such as crankshaft, connecting rods, pistons in combination with a well ported cylinder head which had larger than standard valves. The engine has 11:1 compression, tubular exhaust system, 648 camshaft, high ratio roller rockers and a single 45mm DCOE Weber carburettor.

The carburettor settings are:
38mm choke size
4.5 auxiliary venturis
F2 emulsion tubes
165 main jets
170 air correctors
45 accelerator pump jets
45 F9 idle jet
50 accelerator pump inlet
7.5mm float level shut off height
15mm full droop float setting
Idle screws set at 7/8 of a turn out

Note that the distributor is an electronic one and that the full advance setting is 35 degrees BTDC, while the idle speed advance (at 1500rpm) is 18 degrees BTDC.

This engine has a 'lumpy' idle of 1400rpm, but above 2500rpm smooths out fairly well and will pull exceptionally well from 3500rpm through to 7800rpm. The engine will rev out to 8500rpm, but the real urgency is gone after about 7800rpm and maximum power is at 7200rpm.

Ford 'Pinto' 2000cc SOHC standard engine (Weber)

This standard Sierra engine was a rebuilt engine fitted into a Westfield sports car. The only high performance equipment on the engine is a vernier camshaft pulley and two 40 DCOE sidedraft Weber carburettors.

The carburettor settings are:
130 main jets
190 air correctors
F11 emulsion tubes
35 accelerator pump jets
34mm chokes
4.5 auxiliary venturis
40 F9 idle jets
7.5mm float level shut off height
15mm full droop float setting

The idle advance (vacuum advance disconnected) is set at 12 degrees BTDC and the total advance at 38 degrees BTDC. The vacuum advance is always disconnected when timing an engine and re-connected afterwards so that maximum fuel economy is achieved.

The idle mixture adjusting screws were all turned out to 1 full turn, but finally set at 7/8 of a turn. The CO reading at idle is 2.3%. The engine response is good and the engine pulls through to 6200-6300rpm solidly (power surge falls off immediately at 6300rpm). The average fuel consumption is now 30 to 35 miles per imperial gallon.

Ford 'Pinto' 2000cc SOHC standard engine (Dellorto)

This standard engine is fitted with twin sidedraft Dellorto DHLA (H) carburettors which have fixed size holes on the idle circuit. These carburettors are quite satisfactory for use on such a large four cylinder in-line

engine, even though the idle air bleeds are pre-drilled. 34mm chokes are the smallest that would ever be fitted to such a large standard four cylinder engine (in either 40mm Weber or 40mm Dellorto carburettors).

Note that for all standard 2000cc four cylinder in-line engines fitted with 40mm sidedrafts consider 34mm chokes as being the size to fit.

The carburettor settings are:
34mm chokes
35 accelerator pump jets
140 main jets
180 air correctors
7772.10 emulsion tubes
55 idle jets
7848.3 auxiliary venturis
Standard idle jet holder
1.50 needles and seats
15mm float level shut off height
25mm full droop float setting

The idle adjustment screws required 4 turns out each from the lightly fully seated position (if less turns are used the engine is continually 'popping' at idle, indicating a lean mixture). Because of the fixed nature of the air component of the idle circuit a 55-57 idle jet was used.

In many instances secondhand 40mm sidedrafts are used as opposed to 45mm sidedrafts because they are cheaper and easier to find. Technically 40mm sidedraft carburettors are too small for 2000cc engines because the maximum effective choke size that this sized engine requires is 36mm. That said, these engines in standard form and with limited engine modifications go remarkably well when equipped with 40mm versions of Weber or Dellorto fitted with 34mm chokes. The reason these carburettors work so well on these larger engines is that, due to the rpm range the engine is usually operating in, for 95% of the time they

are providing ideal air/fuel mixture - as good as any larger carburettor could ever supply.

For any two valve per cylinder engine of this size being used up to 6500rpm (and more if the engine has four valves per cylinder) on the road, 40mm sidedraft carburettors will almost always prove to be ideal. The fitting of larger sidedraft carburettors, fitted with larger 36mm or 38mm chokes often does not result in any significant increase in acceleration, although the top end performance will often be improved if the engine is suitably modified. Standard engines do not normally respond to the fitting of larger sidedraft carburettors fitted with larger chokes.

Vauxhall 2000cc 16v engine (Dellorto)

This engine is standard and is fitted into a sports car. The electronic fuel injection has been removed and twin sidedraft DHLA Dellorto 40mm carburettors substituted. The distributor is a Bosch contact breaker points type from the same engine series (bolts straight in) and featured a total advance of 32 degrees BTDC and vacuum advance. This engine has 278 degree duration hydraulic camshafts fitted. Engine performance is excellent.

The carburettor settings are:
34mm chokes
142 main jets
200 air correctors
45 pump jets
50 idle jets
7850.1 idle jet holder
7772.5 emulsion tubes
7848.1 auxiliary venturis
15mm float level shut off height
25mm full droop float setting

The idle adjustment screws were

all set to 3 1/2 turns out from the lightly seated position.

Ford RS 2000cc SOHC Escort (Weber)

The following settings were those required for good all round engine performance. This engine is bored out to 2094cc, has a Group 1 camshaft, Group 1 valves, standard replacement rockers and 10:1 compression ratio. The camshaft is timed 2 degrees retarded (110 degrees).

This engine is revved to 7000rpm only, and is limited to that via the use of a modified 'governor rotor' as fitted to the points type Bosch distributor. The distributor does not have a vacuum advance mechanism (poor economy!). The idle speed (1200rpm) ignition timing is 18 degrees BTDC and the total advance, 'all in' at 3500rpm, is 38 degrees BTDC.

The idle adjustment screws are turned out one complete turn. This engine is very tractable.

The carburettor settings are:
38mm chokes
5.0 auxiliary venturis
F16 emulsion tubes
145 main jets
180 air correctors
40 accelerator pump jets
45F9 idle jets
50 accelerator pump inlet
7.5mm float level shut off height
15mm full droop float setting

Ford Sierra Cosworth 2000cc (naturally aspirated)(Weber)

This engine has a cylinder head which retains the standard valves, enlarged inlet and exhaust ports and L1 camshafts. The engine has 11:1 compression and is equipped with two 45mm sidedraft Weber carburettors.

The distributor is a Bosch points type with the vacuum advance retained. The idle speed advance at 1200 rpm is 20 degrees and the total advance is 32 degrees, 'all in' at 3400rpm. Maximum revs are limited to 7700rpm by means of an electronic cut-out and a mechanical governor rotor.

The carburettor settings are:
F16 emulsion tubes.
38 mm chokes.
5.0 auxiliary venturis.
40 accelerator pumps.
100 accelerator pump inlet.
45 F9 idle jets.
155 mains jets.
210 air correctors.
7.5mm float level shut off height
15.0mm full droop float setting

The idle mixture adjustment screws are set to 1 1/4 turns out from the lightly seated position.

BMC B-Series 1900cc MGB engine (Weber)

Modified engine fitted with a 300 degree duration camshaft, a modified cylinder head and a single 45mm DCOE Weber carburettor.

The carburettor settings are:
36mm chokes
F16 emulsion tubes
55 F9 idle jets
Idle adjustment screws 2 3/4 turns out
50 accelerator pump jets
145 mains
160 air correctors
4.5 auxiliary venturis
7.5mm float level shut off height
15mm full droop float setting

Idle advance is set at 15 degrees BTDC and total advance at 35 degrees BTDC. This amount of idle speed advance is about the limit and,

occasionally, the engine will kick back on starting, especially when cold. The fact is that the engine runs better with this amount of idle speed advance, when the advance is reduced to give better starting, the overall engine performance drops off.

Toyota 4A-GE 1600cc 16-valve MR2 engine (Weber)

This is an otherwise standard engine equipped with a pair of sidedraft 40mm Weber carburettors. This engine is distributor equipped, has 15 degrees of idle speed advance and 32 degrees of total advance and no vacuum advance. Engine runs very well.

The carburettor settings are:
34mm chokes
140 main jets
175 air correctors
35 pump jets
50 F11 idle jets
F 16 emulsion tubes
4.5 auxiliary venturis
No hole in the accelerator pump inlet valve
7.5mm fuel shut off height
15mm full droop float setting

MG 1940cc alloy 8 port Magnette engine (Weber)

This is an older engine prepared for classic racing. It uses a brand new cylinder head, as made by Alexander, and a late model BMC B-Series five main bearing cylinder block, crankshaft and connecting rods.

The carburettor settings are:
38mm chokes
F 11 emulsion tubes
145 main jets
155 air correctors
45 F9 idle jets
50 accelerator pump jets

1 1/4 turns out of the idle screws
7.5mm float level shut off height
15mm full droop float setting

Ford 1600cc Crossflow engine (Weber)

A modified engine using Mexico-sized valves, a well ported cylinder head, 11.5:1 compression, 292 degree duration camshaft and a pair of 40mm DCOE Webers. The spark advance is set at 15 degrees at an idle speed of 1200rpm and there is 35 degrees of total advance 'all in' at 3500rpm.

The carburettor settings are:
34mm chokes
140 main jets
210 air correctors
F16 emulsion tubes
40 accelerator pump jets
4.5 auxiliary venturis
45 F8 idle jets
11/8 turn out of the idle screws
7.5mm float level shut off height
15mm full droop float setting

This is a common engine and a common engine size. The individual settings will vary from engine type to engine type, but the general range of jettings/settings listed are commensurate with the engine size and the degree of modification will give a reasonable guide to what will ultimately be required.

Jaguar XK 3.8 litre engine (Dellorto)

This engine has 8:1 compression and is fitted with 290 degree camshafts, triple 45mm Dellorto carburettors and a well modified cylinder head with 2in diameter inlet valves and standard exhaust valves. The inlet ports were quite large, with the port diameters out to 1.550in/39.5mm up until just

before the valve guide area and then increasing to 1.600in/40.5mm at the valve guide before flaring out to the valve seat. The engine has 12 degrees of idle advance at 1000rpm and 44 degrees of total advance 'all in' at 3000rpm.

The carburettor settings are:
40mm chokes
165 main jets
170 air correctors
35 pump jets
7772.6 emulsion tubes
8011.1 auxiliary venturis
55 idle jets
7850.1 idle jet holder
3 1/2 turns out of the idle adjustment screws
15mm float level shut off height
25mm full droop float setting

Ford 'Pinto' 2100cc engine fitted to a Formula 27 sports car (Dellorto)

This engine is fitted with twin sidedraft 45 mm DHLA Dellorto carburettors. It features a GP1 camshaft and Group 1 inlet and exhaust valves cleaned up inlet ports and well opened out exhaust ports. The compression is 11.5:1, idle speed advance is 18 degrees at 1200rpm and the total amount of ignition advance is 38 degrees 'all in' at 3500rpm.

The carburettor settings are:
38mm chokes
162 main jets
180 air correctors
7772.6 emulsion tubes
8011.1 auxiliary venturis
42 accelerator pump jets
60 idle jets
7850.1 idle jet holders
3 1/2 turns out of the idle screws
15mm float level shut of height
25mm full droop float setting

Ford Sierra Cosworth 2000cc racing engine (Dellorto)

This engine is fitted with 48mm sidedraft Dellortos. It has a fully ported but standard size valve cylinder head, 12:1 compression and BD4 camshafts. The ignition timing is 20 degrees of idle advance at 1500-1600rpm and 32 degrees of total advance 'all in' at 3800rpm.

The carburettor settings are:
40mm chokes
165 main jets
180 air correctors
8011.1 auxiliary venturis
40 accelerator pump jets
7772.6 emulsion tubes
7850.1 idle jet holders
65 idle jets
3 1/2 to 4 turns out on the idle screws
15mm float level shut off height
25mm full droop float setting

Ford 1760cc Crossflow engine (Dellorto)

This engine is equipped with 45mm sidedraft Dellortos. It has a well modified, big valve cylinder head, 310 degree duration high lift camshaft and 11:1 compression. The idle speed spark advance is 15 degrees at 1200rpm and the total amount of spark advance is 35 degrees 'all in' at 3500rpm.

The carburettor settings are:
36mm chokes
142 main jets
190 air correctors
7772.6 emulsion tubes
8011.1 auxiliary venturis
58 idle jets
7580.1 idle jet holders
42 accelerator pump jets
15mm float level shut off height
25mm full drop float setting

BMC/Rover 1275cc A-series engine (Dellorto)

This well modified engine is fitted with a 45mm Dellorto sidedraft carburettor. The ignition advance is 15 degrees at idle and 33 degrees full advance.

The carburettor setting are:
36mm chokes
160 main jets
180 air correctors
7772.6 emulsion tubes
8011.1 auxiliary venturis

40 accelerator pump jets
58 idle jets
7580.1 idle jet holders
15mm float level shut off height
25mm full droop float setting

J.E. Developments
(John Eales)
The Rover V8 engine builder

Cubic Capacity (cc) of available engine sizes

		Bore/mm			
	88.9	**90.0**	**94.0**	**94.5**	**96.0**
63.0			3497.7		
71.1	3531.6	3619.6	3948.5	3990.6	4118.3
77.0			4274.9	4320.5	
80.0			4441.5	4488.8	
79.4					4597.7
82.0			4552.5	4601.1	4748.3
86.3			4794.6	4845.7	5000.8
91.5			5079.9	5134.1	5298.4

(Stroke/mm on left axis)

- Large stock of Rover V8 parts
- Dry sump systems
- Downdraught and side inlet manifolds
- Sump baffles
- Small diameter flywheel and starter motor
- Forged steel adjustable rocker assemblies
- Hewland adaptor plate/bell housing
- Bell housing for T5, Getrag, etc
- Balancing
- Forged steel conrods
- Forged pistons for most capacities at 89.4mm, 90mm, 94mm, 94.5mm, and 96mm
- New cross bolted 3.5 and 3.9 blocks available
- Cross drilled crankshafts at 63mm, 71.12mm, 77mm, 80mm, 82mm, 86.36mm, 90mm and 91.5mm

Claybrooke Mill
Frolesworth Lane
Claybrooke Magna
Nr. Lutterworth
Leicestershire
LE17 5DB

Tel: 01455 202909
email: john@ealesxx.fsnet.co.uk
web: www.rover-v8.com

Index

Accelerator pump 19, 20
Accelerator pump intake/discharge
    valve 24, 25, 76, 77
Accelerator pump piston rods (Weber)
    50
Accelerator pump jet selection 75-77
Accelerator pump jets 19
Accelerator pump jets-final selection
    101, 102
Accelerator pump return springs 19
Acknowledgements 13
Adjusting idle by-pass systems 101
Air correctors 16-18
Air corrector-selection 73, 74
Air filters 60, 61
Air/fuel ratio readings 101, 102
Anti-vibration mountings 82, 83
Auxiliary venturi-selection 74-76

Body components (Dellorto) 41-48
Body components (Weber) 48-58
Butterfly removal 30-32
Butterflies (fitting) 35, 36
Butterfly securing screws 36

Camshaft timing versus idle speed 85

Carburettor strip down (Weber and
    Dellorto) 22-28
Carburettor-checking fit 80-82
Carburettor/s-fitting to inlet manifold
    84-84
Choke location screws 46, 101
Choke size-selecting 65-67
Choke size versus carburettor size 64,
    65
Choke sizes - selecting 65, 67, 103
Chokes and auxiliary venturis 15, 16
Choosing the tuning components for
    your carburettor 64
Cleaning components 22, 23
Clearing passageways 29
Component inspection (Weber &
    Dellorto) 28, 29
Components - initial selection 64-78
CO readings 101, 102

Dellorto accelerator pump 21, 47, 48
Dellorto accelerator pump jets/plugs
    45, 46
Dellorto butterfly fitting 36
Dellorto exploded view diagram &
    parts listing 26, 27

Dellorto floats 21, 42, 43
Dellorto fuel filter 46, 59
Dellorto idle jets and holders 20, 67-
    70, 97, 98
Dellorto idle jet codes 68, 69
Dellorto needle and seat 21
Dellorto progression hole plugs 45
DCO3 carburettors 10, 11
DCO/SP Weber carburettors 50-52
Difficult procedures 29-38

Emission controlled Weber or Dellorto
    carburettors 13, 38-40
Emission controlled Weber
    carburettors (idle jets) 68
Emulsion tubes 16, 19
Emulsion tube selection (Weber/
    Dellorto) 71-73
Essential information 14

Facet fuel pump 63, 107
Fitting butterflies 35-36
Float level setting (Dellorto) 42, 43
Float level setting (brass float Weber)
    49, 50
Float level setting (Spansil float

Weber) 50, 51
Floats, checking condition (Dellorto) 41
Float checking condition (Weber) 48, 49
Float and fulcrum pin (Dellorto) 41, 42
Float and fulcrum pin (Weber) 48, 49
Floats 20, 21, 41
Floats and fulcrum pin fitting 42
Fitting carburettors to inlet manifolds 80-82
Fuel enrichment device 23
Fuel enrichment device, blocking off discharge holes (Weber only) 28
Fuel filters 23, 41, 46, 56, 57, 59
Fuel leakage from fuel enrichment device (Weber) 105, 106
Fuel level & needle valve operation 85, 86
Fuel lines (pipes) & fittings 59, 60
Fuel management 59-63
Fuel pressure 62, 63

Idle by-pass circuitry carburettors 99, 100
Idle jets 18, 19
Idle jet - selecting 67-70
Idle jets 18, 19
Idle jet alteration (fuel component) 67, 68
Idle jet codes (Dellorto) 68, 69
Idle jet codes (Weber) 68
Idle jets/air bleeds (Dellorto) 69, 70
Idle jets/air bleeds (Weber) 69, 70
Idle mixture adjusting screws 16, 18, 44, 45, 55
Idle mixture screw initial adjustment 86-89
Idle mixture adjusting screw - final setting 98, 99
Idle mixture and progression holes (Weber/Dellorto) 70, 71
Idle mixture adjustment 71, 86-89
Idle speeds 85

Inlet manifolds 79-84
Inlet manifold stud alignment 79-82

Jetting examples for a range of engines 109-113

Maintenance 108
Main jets 16-18, 71
Misab spacers 82, 83

Needle and seat 21
Needle and seats (Dellorto) 41
Needle and seats (Weber) 48
Needle valve - selection 77, 78
Norman Seeney Ltd 10, 11

Progression hole inspection 15, 16, 18

Ram tubes 61, 62
Recognizing emission carburettors 38-40
Re-fitting butterflies 35, 36
Removing damaged threaded components 36, 37
Removing jammed chokes and auxiliary venturis 37, 38
Removing jammed throttle spindle nuts 30, 31
Removing butterfly securing screws 31
Removing bearings from carburettor bodies 30-32
Repair kits 10
Reversion plates 62
Return springs 19, 108
Rolling road testing 104
Rubber gromets 82, 83

Solving problems - low and mid-range rpm 104, 105
Solving problems - high rpm 105
Spanish built Webers 99
SP Spanish built Webers 50, 51
Spansil floats (Weber) 50, 51
Spansil float settings (Weber) 51

Spindle bearing maintenance 29, 30
Spindle re-fitting 32-35
Spindle and bearing removal 30-32
Stripping carburettors 23-28
SU fuel pump 62
Synchronizing meters 87, 88, 91, 92

Testing and problem solving 103-106
Testing and setting up 75-102
The importance of stud alignment 79-82
Thackery washers 82
Throttle cable systems 84
Throttle arm fitting 46, 47, 86
Throttle initial settings 86
Throttle setting for single carburettor applications 86
Throttle spindle removal 30-32
Throttle spindle, bearing maintenance 29-32
Throttle spindle and bearing re-fitting 32-35
Throttle syncronization (multiple carburettors) 86
Throttle syncronization 86, 89-93
Top cover (Dellorto) 43, 44
Top cover (Weber) 52, 53
Track testing procedures 104-106
Tuning by exhaust gas CO% 108

Weber accelerator pump intake valve 52, 53, 76, 77
Weber bottom well cover 58
Weber exploded view diagram & part listing 24-25
Weber idle jets and holders 20, 67-70, 96, 97
Weber idle jet codes 68
Weber floats 48-51
Weber return springs 19
Webers - adapting for off-road applications 106-108
Webers - fuel leakage from carburettors 105, 106
Webers versus Dellortos 8-13